T0234351

SpringerBriefs in Applied Sciences and Technology

PoliMI SpringerBriefs

Series Editors

Barbara Pernici, Politecnico di Milano, Milano, Italy

Stefano Della Torre, Politecnico di Milano, Milano, Italy

Bianca M. Colosimo, Politecnico di Milano, Milano, Italy

Tiziano Faravelli, Politecnico di Milano, Milano, Italy

Roberto Paolucci, Politecnico di Milano, Milano, Italy

Silvia Piardi, Politecnico di Milano, Milano, Italy

Springer, in cooperation with Politecnico di Milano, publishes the PoliMI Springer-Briefs, concise summaries of cutting-edge research and practical applications across a wide spectrum of fields. Featuring compact volumes of 50 to 125 (150 as a maximum) pages, the series covers a range of contents from professional to academic in the following research areas carried out at Politecnico:

- Aerospace Engineering
- Bioengineering
- Electrical Engineering
- Energy and Nuclear Science and Technology
- Environmental and Infrastructure Engineering
- Industrial Chemistry and Chemical Engineering
- Information Technology
- Management, Economics and Industrial Engineering
- Materials Engineering
- Mathematical Models and Methods in Engineering
- Mechanical Engineering
- Structural Seismic and Geotechnical Engineering
- Built Environment and Construction Engineering
- Physics
- Design and Technologies
- Urban Planning, Design, and Policy

Stefano Capolongo · Monica Botta ·
Andrea Rebecchi

Editors

Therapeutic Landscape Design

Methods, Design Strategies and New Scientific Approaches

POLITECNICO
MILANO 1863

Editors
Stefano Capolongo
Department of Architecture, Built
Environment and Construction Engineering
(ABC)
Politecnico di Milano
Milan, Italy

Monica Botta
Landscape Architect
Bellinzago Novarese (NO), Italy

Andrea Rebecchi
Department of Architecture, Built
Environment and Construction Engineering
(ABC)
Politecnico di Milano
Milan, Italy

ISSN 2191-530X   ISSN 2191-5318 (electronic)
SpringerBriefs in Applied Sciences and Technology
ISSN 2282-2577   ISSN 2282-2585 (electronic)
PoliMI SpringerBriefs
ISBN 978-3-031-09438-5   ISBN 978-3-031-09439-2 (eBook)
https://doi.org/10.1007/978-3-031-09439-2

This Springer imprint is published by the registered company Springer Nature Switzerland AG
The registered company address is: Gewerbestrasse 11, 6330 Cham, Switzerland

Foreword

After decades of viewing the outdoor space of a healthcare facility as an unimportant addition to the building, the therapeutic significance of these spaces is becoming more and more clear. Design practice and user research is increasingly demonstrating that outdoor spaces in cities and the garden spaces of healthcare facilities can have a significant effect on people's health. In many nations of the developed world, an increasing proportion of older people are developing symptoms of Alzheimer's disease and other forms of dementia. It is laudable that in this volume several chapters focus on designing for this particular patient group, for whom specific design elements can provide a prosthetic environment, thus ensuring peoples' dignity and safety, and alleviating the concerns of the staff.

Children, like people with disabilities and the elderly, are particularly vulnerable to a mismatch between their needs and the design of the physical environment. Italy has experienced an alarming reduction in the proportion of children allowed to travel to school on their own, or to enjoy what has become known as a "free range childhood" in the city. It is particularly important that this concern is being documented and hopefully brought to the attention of city planners and politicians. Children deprived of experiencing a reasonable sense of independence at an early age may have trouble expressing themselves as fully creative adults in later years.

There are many ways of establishing a healthy relationship between human beings and the natural environment. One of these is through horticultural therapy, where benefits for those with physical or mental health problems can be facilitated through the simple act of gardening. Nurturing a plant that needs to be taken care, learning about the value of compost, experiencing a sense of time standing still—all of which can occur in the garden—bring benefits to a patient which may not be possible through medication alone. The design of spaces for horticultural therapy and the inclusion of relevant vegetation require the collaboration of those trained to work with patients together with botanists and ecologists.

In situations where outdoor space is not possible, and the practice of horticultural therapy is not feasible, natural elements can be brought into the environment through biophilic design. Such elements in a building might include views to greenery, indoor plants, adequate daylight to enhance circadian rhythms, nature-related artwork, and architectural design details. Close collaboration must be established between architects, landscape architects, and interior designers to ensure the best possible solutions and outcomes of biophilic design.

Overall, this volume presents a significant collection of research findings and relevant design directives. The design of buildings and outdoor spaces undoubtedly affect the day-to-day lives and health of those who use them. The publication of this important book presents research-grounded information that will enable readers to influence, design, and build environments beneficial to those who live, work or play in them.

Clare Cooper Marcus
Professor Emerita
Departments of Architecture
and Landscape Architecture
University of California
Berkeley, USA

Preface

The health status—as considered today—is not just an individual protection and promotion issue, but a collective condition, strongly influenced by the environmental context; the link between the morphological and functional features of urban contexts and places, and the Public Health outcomes, opening up to a new scientific and design scenario about Urban Health research topic.

Referring to this research field, the introduction of Nature-Based Solutions and Therapeutic Landscape Design approaches into the healthcare contexts, facilities, and architectures has assumed a considerable importance over the years. The recent period that saw the healthcare facilities fighting against the COVID-19 pandemic, highlighted the need, especially in complex periods of lockdown, to use green spaces to recover health and well-being both in hospitals and in public or private places. The COVID-19 pandemic is an important demonstration of the dual effects of urbanization on the environment, that is, the intrinsic capacity of the contemporary city to be a place of economic and social opportunities and, at the same time, places where multiple risk factors for Public Health and Health Welfare could be developed and disseminated.

Starting from the experience developed into the six editions of the Training Course in "Therapeutic Landscape Design"—which has been held since 2015 at the Department of Architecture, Built environment and Construction engineering (ABC) of Politecnico di Milano, where the multidisciplinary approach of the Lectures involved helps to provide scientific, technical, healthcare, social and design methodologies to over 120 attendees—the need to bring together, in a scientific monograph, the contribution of the Professors, Professions, and Practitioners, became crucial as much as relevant.

A more conscious design of therapeutic green spaces—especially referred to the Healing Gardens—can give texture and consistency to several unique elements in the treatment of specific pathologies, to support social discomfort, to develop healing processes and practices, and to improve the evolution of healthy lifestyle and accessible built environment for fragile users and people with disabilities.

This collection of insights and scientific experiences aims to provide the tools to plan and create therapeutic places and natural spaces aimed to support the care process, as well as giving an overview of national and international case studies, defining design approaches, analysis, and best practices.

Milan, Italy   Stefano Capolongo
                                                                                               Monica Botta
                                                                                          Andrea Rebecchi

Contents

Urban Health: Applying Therapeutic Landscape Design. Methods, Design Strategies and New Scientific Approaches

Gayle Souter-Brown

1 Research Outlook

Recent increased urbanisation and urban densification, coupled with corporate agriculture and horticulture, have changed the environment. Concurrently, digital lifestyles have disconnected people from nature. Stress levels are rising. Cities are increasingly impermeable, new developments often lack private gardens and public parks are shrinking. Young and old people spend less time outside as there is little to attract them outdoors. Although overwhelming evidence links positive outcomes for health and well-being with the stress reducing properties of the environment, a dose of nature, through views and time spent outdoors, is increasingly difficult to find. New research seeks to integrate theory and practice.

Therapeutic landscapes have been a feature of human settlement since ancient times. Intentionally richly planted, they offer serenity, contact with nature, space and a sense of refuge. Contemporary therapeutic landscapes are generally developed as gardens, on a human scale.

2 Why Do Towns and Cities Need Therapeutic Landscapes?

The health and wellbeing of people and planet is intimately linked. As urbanization increases, so environmental pressures and human stress levels increase. Rising chronic stress levels across towns and cities affect community vitality. To ease the effect of these pressures, a new balance must be sought whereby the therapeutic properties of landscapes are prioritized within urban planning and design. Human scale landscapes are at the heart of healthy cities.

G. Souter-Brown (✉)
Auckland University of Technology, Auckland, New Zealand
e-mail: gayle@greenstonedesign.co.uk

Historically, urban health was supplied with biodiverse tree-filled public parks and gardens providing clean air and somewhere to take gentle exercise. With reduced public budgets and an ongoing housing crisis many of these former 'lungs' of the city have been sold off to developers or neglected.

The recent global Covid-19 pandemic places new attention on health promotion as vital to ensure health systems are not overwhelmed. The protective nature of reduced stress and enhanced community wellbeing must be recognised in design research and practice. Therapeutic landscapes, supplied as a network across all parts of our towns and cities, address a need for mental health support, reducing stress to prompt and enable healthy lifestyles. Protecting existing greenspace, promoting new therapeutic landscapes and enhancing the health-giving qualities of urban nature will not only aid human health, but address issues around climate change, biodiversity loss, and aid economic recovery.

3 Evidence Base

The evidence base for the efficacy of therapeutic landscapes is significant [4–10]. Interest from mental and physical health, landscape and urban design professionals is growing, alongside diverse groups such as sociologists, urban foresters, housing providers and policy makers [11, 12].

3.1 Biophilia

E.O Wilson's theory of biophilia [1], our innate love of living things, is fast becoming a key instrument in fashioning functional urban design. Taking an evolutionary approach, Wilson found that humans are wired to respond positively to nature contact.

3.2 Attention Restoration Theory

Kaplan's Attention Restoration Theory [2] takes the biophilia hypothesis further and posits that time spent in nature is protective; it promotes and enhances health and well-being.

3.3 Salutogenesis

A shift towards a focus on health promotion and prevention of illness is captured by Aaron Anonovsky's term salutogenesis [3]. Salutogenesis offers an alternative to traditional pathogenic healthcare and invokes design to function as an effective public health promotion tool.

Salutogenic design invokes biophilia, brings nature into the city and thus engages landscape architects as public health professionals. This is significant because it is widely accepted that prevention is more cost efficient than mitigating the effect of such wide-ranging challenges as disease, ageing and climate change.

4 Putting Theory Into Practice

Unequivocal research reports that nature and natural scenes are vital for health and well-being. However, too many children and adults cannot access nature in their daily lives. Some have no 'lived experience' of a tree. The current design trend towards straight lines and away from the natural variety of nature ignores research is both a public health challenge and an opportunity to reshape the way we view urban environments.

5 Design Guidelines

As early as 1960, Harold Searles realized "The nonhuman environment, far from being of little or no account to human personality development, constitutes one of the most basically important ingredients of human existence" [13]. When we design for health and well-being it is essential that we remember this.

Therapeutic landscape design is a mix of art and science. To answer the question of how to design a therapeutic garden, it is helpful to investigate the past. The first 'therapeutic gardens' can be found in the Islamic and monastic religions. Realizing nature's capacity to inspire, to give peace, to heal and to balance life, early gardens were built to sustain the local community as well as to aid health and wellbeing in a time of pestilence and strife. They followed the imagination of 'paradise', with a sense of mystery and a deep, spiritual appeal. The early therapeutic landscapes shared four primary features: water and shade, symmetry, edible and sensory planting, and serenity. The people's search for peace made the garden aim to be a beautiful sanctuary from the world outside, a place for calm repose. In Arabia, where the Islamic religion originated, water was a precious commodity. Its availability impacted the choice and placement of plants. Symmetry was deliberately created by repetitive planting. As edible and sensory planting was important, almonds were planted near seating spaces and fragrant citrus fruits near dining areas. To prevent soil moisture evaporation, underplanting with plants such as lilies and roses was necessary. A sense of peace was found in soothing Islamic gardens. To counter the arid world beyond the garden walls, they needed to be green, secluded spaces, which emphasized the contemplative dimension. A balance of water, planting and symmetry allowed a harmonious connection to nature. In addition, edible planting created a sensory experience as people could touch, smell and taste the fruits. Fruit trees also provided welcome shade while blossoms attracting bees and insects provided colours and scent.

Monastic and Islamic gardens shared many characteristics.

Monastic gardens aimed to be both practical and beautiful. They can be divided in four main functional areas: The kitchen garden provided some cheap and fresh fruits and was built by lay priests working to be able to enjoy the welfare and protection of the church. Thus, it was a kind of early 'community garden'. Reducing the risk of pests, flowers were planted between the vegetables. Next, the orchard offered fruit and nuts as well as hop vines for medicinal use and beer. In the third part, the physic garden, medicinal herbs were planted. Following the goal to identify the herbs easily, they built one bed for each herb. The cloister garden as last part was used as a contemplative space, consisting of a level field of lawn surrounded by fragrant planting. The church was expected to feed the poor and heal the sick and to meet those expectations with minimal elements, as shown in Fig. 1. The gardens they created thus accommodated both basic physical and spiritual needs.

Like monastic gardens, Islamic gardens were divided into four by the four paradise rivers of the Quran believed necessary to fulfil the need of creative, organized thought. As edible and sensory planting was important, fragrant, tasty fruit and nut trees were planted near seating spaces and citrus fruits near dining areas. To prevent soil moisture evaporation, underplanting with lilies and roses was used. Keeping in mind the belief in paradise on earth, intimate, compact and welcoming spaces removed from the dirty, noisy world outside, were built in form of gardens. These inspired the senses, with health-giving herbs, clean air and views to heal. These basic needs remain the same

Fig. 1 Cloister garden with elements such as seats and planting for physical and spiritual needs. Photo credit: Author

today. The universal concepts of peace and unity of the historic Islamic gardens are still used. They were spaces to feed mind, body and spirit, provide a sense of safety and enclosure. Never created in isolation, they were built in working and living environments. One can learn from Islamic-style planting and the careful use of water as cities adapt landscape and urban design to climate change. Importantly, sustainable therapeutic landscapes positively impact public mental health.

6 Applying Biophilic Design

As urbanization has increased, the urban environment has become increasingly degraded, and the incidence of lifestyle-related diseases grown. At the same time there has been increasing interest in how urban ecology, architecture, socio-economic and academic/work outcomes, health and well-being intersect. To date, researchers have used a relatively narrow, discipline-defined lens to examine potential linkages. The socio-psychologist Eric Fromm's theories of personality first raised the term "biophilia as a potential cue for many innate behaviors (Fromm, 1964). The ecologist Edward O. Wilson took the idea further, to propose the biophilia hypothesis. In his book Biophilia, he stated that "our natural affinity for life—biophilia—is the very essence of our humanity and binds us to all other living things" [14]. This approach asserts that humans have an innate connection with nature that can assist to make the urban environment more effective, supportive human abodes. In an urban context, opportunities to connect with nature can be problematic. For the purposes of this essay then we offer "therapeutic landscapes", "gardens", and "environmental design" as a means to facilitate the necessary nature connection within an urban setting [15]. Biophilic design is thus articulated by the design profession as the relationships between nature, human biology, and the built environment [16]. Biophilic design brings nature contact into every day experience; it offers opportunities to connect with nature, which is essential for a life in a healthy place and community.

Therapeutic landscapes are so-called because they create a physiological effect on the people who use them. People of all ages are attracted and respond to a biophilic landscape. The opportunity to engage with urban nature provided through a biophilic design approach is what tempts children and adults who characteristically spend their time indoors to venture outside, to receive the health and well-being benefits of nature connection. The student in Fig. 2 had never previously climbed a tree, but when presented with the opportunity within a therapeutic garden setting was naturally motivated to explore its heights. For many people the response is an unconscious desire to connect with nature. Some previously inactive people may find themselves spontaneously more physically active, eating better, feeling more productive. When people are encouraged and enabled to spend time outdoors, their risk of developing lifestyle-related health conditions such as heart disease and some cancers is reduced. Similarly, in the older population, dementia and depression risk is reduced [17, 18].

Biophilic design has diverse applications. Roger Ulrich showed a positive impact of natural landscape on rehabilitation. Examining patients post-surgery with different

Fig. 2 Student climbing a
tree in a therapeutic garden.
Photo credit: Author, with
permission from the subject

window views (one into nature, the other one into a brick-built wall) he found that "the patients with the tree view had shorter postoperative hospital stays, had fewer negative evaluative comments from nurses, took fewer moderate and strong analgesic doses, and had slightly lower scores for minor postsurgical complications" [19]. Although he voiced misgivings that these results apply to every disorder, he recommended "hospital design and siting decisions should consider the quality of patient window views".

Stephen Kellert describes the goal of biophilic design as creating places imbued with positive emotional experience that are the precursors of human attachment to and caring for place (49).

The positive emotional experience is an important feature of a therapeutic landscape. In Fig. 3 the photo shows a new playground, installed by a commercial play provider as part of the British Labour government's national Play Strategy. Lacking reference to therapeutic landscape design theory, although set in the middle of a housing development, children actively bypass the playground on their way home from school as it offers less interest than the open space and surrounding boundary planting. The playground shown lacks biophilic elements, in short, life. As Heerwagen says "Not all nature [or green space] is equally attractive or beneficial. Spaces with dead and dying plants and trees signal habitat depletion and are largely avoided. In contrast, places with rich vegetation, flowers, large trees, water, and meandering pathways … are sought out … as places of relaxation and enjoyment" [20].

Fig. 3 Unattractive playground installed as part of a previous British government's national Play Strategy. Photo credit: Author

Biophilic landscape design impacts health and wellbeing. To achieve positive health outcomes the design needs to be mindful of:

- nature connection opportunities, such as planting for seasonal change, vistas and sky, and include environmental features such as water, wildlife-attracting plants, natural materials and colors;
- variety, such as trees, flowers and animals, create information richness, spirituality, attraction and arouses curiosity;
- a balanced mix of order and complexity, as too much complexity affects the curiosity negatively. People should desire to know more about a space with increased exploration;
- natural processes, birth, growth and death should be included;
- opportunities to use recycled elements as the natural aging of materials create an impression of resilience in the built environment [21];
- a sense of refuge, safety and protection, necessary for a relaxing experience;
- cultural or historical features like sculpture, edible or native planting helps people to connect to the place. "Using inspiration from both the local natural environment and vernacular cultural expressions for creating a sense of place is critical to the success of biophilic design" [20];
- Heraclitan motion as meaningful aspect, as nature is always on the move and some kinds of movement patterns may be associated with safety and tranquility (soft movement that always changes, trees or grass in a light breeze), while others indicate danger through erratic movement changes indicating wind, storm. ibid.

We know ecological health is required for human health and well-being. Thus, correctly applied, biophilic design can create a therapeutic landscape that is ecologically balanced, with friable soils and a range of beneficial wildlife-attracting planting. In summary, a therapeutic landscape provides the necessary habitat for life, for people and planet—clean water, fresh air, sunshine and shelter. Trees, shrubs, flowers and ground covers decorate the space, offering beauty, fragrance, colour, texture, taste and sound. In areas where predatory animals are a problem, habitat must be adjusted to ensure human safety. However, birds, fish, insects, invertebrates, small reptiles must be made welcome, through appropriate environmental management techniques. There is no place for inorganic pesticides and herbicides in a therapeutic landscape.

7   Applying Salutogenic Design

The "salutogenic model", as a theory to guide health promotion, posits that it is better, and less expensive, to prevent disease, to address the social determinants of health [22]. Utilising a sense of coherence, a salutogenic methodology provides accessible nature connection points as health supports within the community, to promote and enhance well-being, to reduce stress, to prevent young and old from becoming unwell. The Landscape Institute state that "throughout history landscape architecture can be linked to the need to create places that were beneficial for people's health and well-being" [23]. Mental health is closely linked to physical health [24]. If we focus on physical health alone we miss a key driver for overall well-being. Architecture recognizes the potential health impacts of design. Ecologists concerned at the environmental implications of a population disconnected from nature believe the growing demand for human well-being could provide environmental benefits [25, 26]. Nature connections, whether through forest walking or urban landscape design interventions, have been shown to reduce stress [27]. Stress is a primary prompt for mental and physical illness. Hence, a salutogenic design approach could be a powerful tool for health and well-being.

Salutogenic design aims to:

1. Prevent illness through promotion of an active healthy lifestyle using natural sensory rich environments
2. Soften the built environment and work with it to create holistic environments that maximize potential of the site, the budget and community well-being
3. Promote wellness and reduce rehabilitation times post trauma and infection, improve mobility, memory, mood, reduce aggression, stress and requirement for chemical pain relief, improve outcomes for diabetes, obesity, heart and lung disease, some cancers and depression, using attractive, engaging soft landscape treatments.

The salutogenic approach to health and well-being is effective because of biophilia "…We respond particularly positively to grassland, trees, edible plants and friendly animals... Therapeutic gardens then must include those elements ... in order to achieve best value and maximum effect" [15].

Nature's complexity and simplicity, as shown in Fig. 4, are important within therapeutic garden design. Experience with wildlife, such as listening to bird song or watching butterlfies in flight allow people to concentrate on nature and forget their daily stress for a moment. Therapeutic gardens provide a reference point. That includes cultural reference to create a sense of place as well as personal reference to create a sense of belonging. Art, sculpture, and native planting connect people to the area and promote feeling of comfort.

As such, the principles of good garden design, texture, height variation, colour and scent, are applied. General features of therapeutic gardens designed using a salutogenic approach are:

- attractive environment to look at from indoors, as well as to be in outdoors—offer a reason to get outdoors;
- functional, so that healthy, active, meaningful activities may be lead in the space
- practical, so that maintenance costs are minimised;
- cost-effective, designed as a beneficial therapeutic environment (to reduce costs of care/intervention in school/recidivist behaviours);

Fig. 4 Therapeutic garden, as could be developed in a residential, workplace, health or educational setting. Photo credit: Author

- balanced, so that the overall space affords rest as well as activity.

More specifically, a mix of light and shade is important. People benefit from being outdoors under natural light. Sunshine is necessary for health and vitality. Feelings of being warm in the sun and cool in the shade enhance the sensory experience. With sunshine, Vitamin D is absorbed, positive mood and energy levels are increased, biological processes enhanced and conditions such as myopia and rickets prevented. More natural light correlates with a healthier life. Different types of shade-giving features such as broad canopy trees or pergolas, create different light levels. Variations in light and shade allow sight-impaired and other sensory-impaired or developmentally delayed individuals to experience their environment more deeply and bring colours to life.

Salutogenic landscapes are engaging. They have flexible, comfortable seating areas, which offer possibilities to just sit, close the eyes, forget the surroundings and enjoy the time, alone or in a group. It is not important whether people sit on high park benches, on the grass or in bean bags, it is the opportunity for varied and variable seating that is essential.

Another important aspect within the therapeutic landscape sensory experience is textural detail. When people walk barefoot, they can feel the change of uneven surfaces. For children, it is important to experience uneven surfaces before the age of 7 to develop balance and proprioception skills. Barefoot paths offering sections of changing surface materials, for example mulch, sand, cobbles and bricks, are a simple, effective method people use to massage their bodies for improved health and well-being. Furthermore, surfacing choices offer possibilities for sustainable urban drainage. Using permeable materials combines drainage with different surfaces to walk on. However, designed landscapes must offer an option for smoother paths as well, for example for people using wheelchairs or bikes, as uneven surfaces are sometimes not productive. Water is a vital feature within healthy cities as it is essential to all life. There are many options to include water into a garden, park or plaza space. It can be temporary or permanent, standing or moving. While some features must remain constant, therapeutic landscapes change and improve over time, as trees grow and wildlife visits or moves in. Accessibility and a commitment to inclusion is a constant requirement and of prime importance. People of all ages and physical and mental states must feel secure to find and move about the place safely. There is no sense of the therapeutic quality within a garden unless everyone can access it.

As the principal of a large school said "By adding seating and softening the play we hope to reconnect children with their imaginations, sense of curiosity, exploration and adventure. We hope their social skills will improve and attention span increase and that these benefits will flow through to the classroom and onwards into society" [15].

8 Conclusion

Therapeutic landscapes integrate theory with practice to offer a cost-effective means to address social, economic and environmental challenges. Intuition and ancient wisdom have been researched and added to, with a compelling evidence base now in place. Unequivocal research shows real world experience of therapeutic landscapes must be made available to people everywhere for their health and well-being to flourish. Health, education and housing providers must match innovative employers to offer therapeutic landscapes as part of their site development. The health of the planet and that of the people is intrinsically linked. When we create a therapeutic landscape, everyone wins.

As urbanization has increased, communities sought to establish cultural contact with local contexts, landscapes, and values. They looked to architects to design buildings and places to facilitate this. Therapeutic landscapes are a way to save money, reduce hospital waiting lists, boost wellbeing, prevent dementia and depression. Reducing stress, improving cognitive function, and expediting therapeutic, healing landscapes help people to live a healthier life. With continuing urbanization those aspects are ever more important [28].

References

1. Wilson EO (1984) Biophilia: the human bond with other species. Harvard University Press, New York
2. Kaplan S (1995) The restorative benefits of nature: toward an integrative framework. Environmental Psychology 15(3):169–182
3. Antonovsky A (1996) The salutogenic model as a theory to guide health. Health Promot Int:11–18
4. Kaplan R (1995) The urban forest as a source of psychological well-being In Bradley GA (ed) Urban forest landscapes: integrating multidisciplinary perspectives. Seattle University of Washington Press
5. Roe J, Aspinall P (2011) The restorative benefits of walking in urban and rural settings in adults with good and poor mental health. Health Place 17(1):103–113. https://doi.org/10.1016/j.healthplace.2010.09.003
6. Shanahan DF, Bush R, Gaston KJ, Lin BB, Dean J, Barber E, Fuller RA (2016) Health benefits from nature experiences depend on dose. Sci Rep 6:28551. https://doi.org/10.1038/srep28551
7. Soga M, Gaston K, Yamaura Y (2016) Gardening is beneficial for health: a meta-analysis. Preventive Medicine Reports. https://doi.org/10.1016/j.pmedr.2016.11.007
8. Ulrich RS, Berr LL, Quan X, Parish JT (2010) A conceptual framework for the domain of evidence-based design. HERD Health Environ Res Design J 4(1):95–114. https://doi.org/10.1177/193758671000400107
9. Velarde MD, Fry G, Tveit M (2007) Health effects of viewing landscapes – landscape types in environmental psychology. Urban Forestry Urban Greening 6:199–212
10. Ward Thompson C, Roe J, Aspinall P, Mitchell R, Clowd A, Millere D (2012) More green space is linked to less stress in deprived communities: evidence from salivary cortisol patterns. Landsc Urban Plan 105(3):221–229
11. Von Lindern E, Lymeus F, Harting T (2016) The restorative environment: a complementary concept for salutogenesis studies. In: Mittelmark MB, Eriksson M et al (ed) The handbook of salutogenesis. Cham (CH): Springer. https://doi.org/10.1007/978-3-319-04600-6_19

12. World Health Organization Europe (2016) Urban green spaces: a review of the evidence. Accessed from http://www.euro.who.int/__data/assets/pdf_file/0005/321971/Urban-green-spaces-and-health-review-evidence.pdf?ua=1

13. Searles H (1960) The non-human environment: in normal development and schizophrenia. International Universities Press Inc., New York

14. Wilson EO (1986) Biophilia New York: Harvard University Press

15. Souter-Brown G (2015) Landscape and urban design for health and well-being: using healing, sensory and therapeutic gardens. Routledge Press, London, England

16. Browning WD, Ryan CO, Clancy JO (2014) Patterns of biophilic design: improving health & well-being in the built environment. Terrapin Bright Green, LLC, New York

17. Dinas PC, Koutedakis Y, Flouris AD (2011) Effects of exercise and physical activity on depression. Int J Med Sci 180(2):319–325

18. Gonzalez MT, Kirkevold M (2014) Benefits of sensory garden and horticultural activities in dementia care: a modified scoping review. Clinical Nurs 23(19–20):2698–2715

19. Ulrich RS (1984) View through a window may influence recovery from surgery. Science 224(4647):420–421

20. Heerwagen J (2009) Biophilia, health, and well-being. In: Campbell L, Wiesen A, (ed) Restorative commons: creating health and well-being through urban landscapes. Northern Research Station: U.S. Department of Agriculture, Forest Service; p 38–57

21. Krebs CJ (1985) Ecology: the experimental analysis of distribution and abundance. Third, Edition. Harper and Row, New York

22. Mittelmark MB, Bull T (2013) The salutogenic model of health in health promotion research. Glob Health Promot 20(2):30–38

23. Landscape Institute (2015) Public Health. Be a Landscape Architect

24. Canadian Mental Health Association (2016) Connection between mental and physical health Ontario: Canadian Mental Health Association. http://ontario.cmha.ca/mental-health/connection-between-mental-and-physical-health/

25. Sadler BL, Berry LL, Guenther R, Hamilton DK, Hessler FA, Merritt C, et al (2011) Fable hospital 2.0: the business case for building better health care facilities. Hastings Cent Rep 41(1): 13–23

26. The University of Exeter's Environment and Sustainability Institute. European Centre for Environment and Human Health. University of Exeter (2015)

27. Capaldi CA, Passmore H-A, Nisbet EK, Zelenski JM, Dopko RL (2015) Flourishing in nature: a review of the benefits of connecting with nature and its application as a wellbeing intervention. Int J Wellbeing 5(4):1–16

28. Griffin MI (2013) Playground features to improve social skills and attention span. In: Souter-Brown G (ed) Landscape and urban design for health and well-being. Routledge Press, London, p 224

Biophilic Design: Nine Ways to Enhance Physical and Psychological Health and Wellbeing in Our Built Environments

Bettina Bolten and Giuseppe Barbiero

1 Introduction

Biophilic Design is an applied science, aimed at planning artificial spaces that reflect the innate tendency of human beings to seek connections with Nature. It is well known that the application of Biophilic Design reduces stress, stimulates creativity and clear thinking, improves physical and psychological wellbeing and accelerates healing (for a review, see [3]).

2 Biophilia

Biophilia is "the innately emotional affiliation of human beings to other living organisms" [34]. It covers a variety of attitudes [27], emotions [6] and values [22] which, collectively, constitute our relationship with Nature.

2.1 Biophilia and Biophobia

According to E.O. Wilson, "biophilia is not a single instinct but a complex of learning rules that can be teased apart and analyzed individually. The feelings molded by the learning rules fall along several emotional spectra: from attraction to aversion" [34]. Attraction is biophilia, aversion is biophobia [32]. During evolution, humankind had

B. Bolten (✉) · G. Barbiero
The Laboratory of Affective Ecology (LEAF), University of Valle d'Aosta, Aosta, Italy
e-mail: bettina.bolten@hotmail.com

G. Barbiero
e-mail: g.barbiero@univda.it

© The Author(s), under exclusive license to Springer Nature Switzerland AG 2023
S. Capolongo et al. (eds.), *Therapeutic Landscape Design*,
PoliMI SpringerBriefs, https://doi.org/10.1007/978-3-031-09439-2_2

to face the hostile forces of Nature in wilderness environments. The learning rules of biophilia and biophobia rooted themselves in the genetic heritage of our species, according to the contribution they made to improving human efficiency in seeking resources and refuges. Wilderness environments trigger two types of physiological reaction: (1) the 'fight-or-flight' response, which translates into a hyperactivity of one of the branches of the autonomic nervous system, usually the over-expression of the sympathetic nervous system [30], which was linked to the concept of biophobia (e.g. [32]) and (2) the 'rest-and-digest' response, which manifests as the cooperation of both branches of the autonomic nervous system, with a prevalent influence of the parasympathetic nervous system. This assures better long-term resilience of the individual [19], as it reduces stress [32] and enhances cognitive functions [21]. Although various scholars consider biophobia to be part of the biophilic system (e.g. [32, 33]), for the purposes of studying Biophilic Design, it would be more convenient to maintain a distinction between the two concepts of biophobia and biophilia [4]. A reasonable objective of Biophilic Design could be to construct environments that can stimulate biophilia [2] and reduce the stress induced by bio-phobia: in other words, environments that can sustain and improve the equilibrium of the autonomic nervous system.

2.2 An Evolutionary History of Biophilia

Biophilia developed in the Palaeolithic period. For approximately 95% of their evolutionary history, human beings survived by adopting a nomadic hunter-gatherer lifestyle. Humans have thus perfected a set of responses adapted to the various wilderness environments—mainly the savannah [29]—aimed at recognizing the quality of an environment in terms of resources and refuges. Some of the environmental preferences which incorporated into Biophilic Design are based on innate learning rules derived from our ancestors' survival, and even today they form the primary, deepest core of our biophilia [13]. After farming was invented, approximately 14,000 years ago [1], most of the human population became sedentary. Human beings started to distinguish the domestic from the wilderness environment. Their refuges became permanent, and the first human clusters were formed: villages and then towns and cities [18]. In this period, which covers approximately 5% of the evolutionary history of humankind, the biophilia structured in the Palaeolithic period was adapted to the new cultural requirements. One example is proxemics. In the Palaeolithic period, groups of *Homo sapiens* were few, and meetings between humans were rare, outside of their own clan. During the Neolithic period, village life required a level of socialization that imposed a hitherto unknown physical proximity, to which we have never fully adapted. This explains, for example, why many people seek outdoor spaces in Nature in which the human presence is rare. Finally, over the past 250 years—an irrelevant period from an evolutionary point of view: less than 0.2% of the evolutionary history of humankind—human beings developed their inclination to trans-form their environment permanently and irreversibly [17]. During this period, human clusters gradually became larger and denser. Compared to the wilderness environments in which humans evolved, towns and cities—now home to 53% of the world's population [35]—are characterized by a lack of green spaces, large crowds, and artificial lighting [7]. The lack of natural stimuli atrophied biophilia [12, 35]. After the industrial revolution, detachment from Nature became even more pronounced. This detachment was so hard that many people feel the need to restore their biophilia by immersing themselves in Nature during their free time.

3   From Biophilia to Biophilic Design

"Biophilic Design is the deliberate attempt to translate an understanding of the inherent human affinity to affiliate with natural systems and processes—known as biophilia—into the design of the built environment" [23]. "Biophilic Design is not about greening our buildings or simply increasing their aesthetic appeal through inserting trees and shrubs. Much more, it is about humanity's place in nature, and the natural world's place in human society, a space where mutuality, respect, and enriching relation can and should exist at all levels and emerge as a norm rather than the exception" [25]. These definitions come from Stephen R. Kellert (1943–2016), Tweedy/Ordway Professor of Social Ecology at Yale University. Kellert, together with E.O. Wilson, is the author of *Biophilia Hypothesis* [27]. Like Wilson, Kellert is also an ecologist, who gradually developed an interest in artificial environments, culminating in the book *Biophilic Design* in which [26] collected the experiences of biologists, psychologists and architects joined by their common interest in artificial environments that respect human biophilia. The first chapter of this book [23] continues to be a reference work for studies on Biophilic Design even today [31].

3.1   Design by Nature: The Legacy of Stephen Kellert

The goal of Biophilic Design is only apparently simple. Kellert saw two limitations that hamper the introduction of effective Biophilic Design: "the limitations of our understanding of the biology of the human inclination to attach value to Nature, and the limitations of our ability to transfer this understanding into specific approaches for designing the built environment" [23]. Therefore, Kellert recognized two dimensions of Biophilic Design. The first was an *organic* or *naturalistic dimension*, inspired by biophilia that established itself genetically during the Palaeolithic period. The second was a *vernacular dimension*, which developed after the Neolithic period. Kellert correlated these two dimensions to 72 'attributes' of Biophilic Design and systematized them according to six 'elements': Environmental features, Natural shapes and forms, Natural patterns and processes, Light and space, Place-based relationships, and Evolved human-nature relationships [23]. The 72 attributes provided a foundation for the *Biophilic Quality Index* by Berto and Barbiero [13].

Kellert's research was interrupted prematurely in 2016. In the book *Nature by Design* [24], published posthumously by his wife Cilla, Kellert sought to systematize Biophilic Design according to three categories: Direct Experience of Nature; Indirect Experience of Nature, and Experience of Space and Place, which led to a series of suggestions aimed at helping designers to incorporate the human affinity with Nature into the built environment. If used appropriately and specifically, instead of as a checklist applied indiscriminately, these suggestions, or alternatively the 72 attributes published in [23], offer a series of options for using Biophilic Design in an effective way [24].

3.2   The 15 Patterns of Biophilic Design by Terrapin Bright Green

A pragmatic approach to Biophilic Design has been proposed by the consulting firm Terrapin Bright Green (TBG), founded by Bill Browning and Cook&Fox Architects. TBG's proposal is based on a systematic collation of environmental psychology literature, concerning the effects of the built environment on human beings. TBG's aim was to identify patterns which have both a scientific foundation and a feasible application by architects in Biophilic Design [14, 15]. Particularly significant is the fact that the entire 'Nature of the space' dimension—which includes patterns 11 to 15—raises the issue of considering, within Biophilic Design, environments that can support and improve the equilibrium of the autonomic nervous system which, as we have seen, is the biological root of biophilia. The 15th "awe" pattern was recently added by [15] to the original 14 models [14]. 'Awe' can be traced back to the penultimate of Kellert's 72 attributes, 'Fear and awe' [23]. This is an item that still need to be improved, including the related Immanuel Kant's meaning of 'sublime' ("das Erhabene" [20]).

3.3   Thirteen Years of Biophilic Design Theories: A Comparative Analysis

We compared the features of Biophilic Design described in the most scientifically relevant publications [14, 15, 23, 24] in order to identify the issues that the authors unanimously considered to be basic to Biophilic Design (Table 1). We noted that the first four attributes (Light; Protection and Control; Air; Views) are considered in Evolutionary Psychology to be essential in the search for refuge, while the next three (Greenery; Curiosity; Biophilic Materials) are essential in the search for resources. It is not surprising that the characteristics of Biophilic Design considered to be universal follow the adaptive models that were developed by our species in its search for a habitat with reliable refuges and resources. It is also unsurprising that the top places

Table 1 Comparison of the most important features of Biophilic Design according to the most relevant studies. The final column on the right contains a summary of our proposal

Kellert [8]	Browning et al. [12]	Kellert [8]	Our summary
Natural light	Dynamic and diffuse light	Natural light	Light
Prospect and refuge	Prospect and refuge	Prospect and refuge	Protection and Control
Air	Thermal and airflow variability	Air	Air
Views and vistas	Visual connection with nature	Views	Views
Plants	Visual connection with nature	Plants	Greenery
Curiosity and enticement	Mystery	–	Curiosity
Natural materials	Material connection with nature	Materials	Biophilic Materials
–	–	–	Sounds
–	–	–	Smells

are held by the issues most closely linked to our biology (the senses), and the cultural, experiential issues are lower down. Finally, we were quite amazed to note that issue of 'Quiet' is never considered. In our view, this issue would deserve greater attention [5, 8]. Since biophilia depends on the balance of the autonomic nervous system, we believe it is important to add two categories related to the 'Quiet' issue: Sounds and Smells. Sounds and Smells seem to be good indicators of a rest-and-digest state [28] and can be part of both finding refuge and seeking resources [16].

4 The Future of the Biophilic Design

In the future, empirical attempts to test Biophilic Design 'in the field', as has happened in recent years, will no longer be sufficient (for a review, see [24]). We think that there is a need to go beyond the list of 'suggestions for designers' on what is important for proper Biophilic Design [24]. The aim of Biophilic Design is to design artificial environments that have a positive effect on individual health and wellbeing. These positive effects need to be measurable. To guarantee that the biophilic quality of projects continues to improve, in the future we will need to establish guidelines derived directly from the results of ap-propriate tests, conducted according to scientific criteria. In the next phase, these guidelines could then be converted into a handbook to assist designers in ensuring the success of their work, and this could be personalized and optimized for each specific case. Finally, in our view it is important to reconnect human beings with Nature [24] rather than "bringing nature into the built space" [14]. The practice of Biophilic Design touches on very deep parts of the human psyche, which are linked to the need, felt by many people, to rediscover an affinity with Nature and feel at one with it again [4]. This also entails an acceptance of the dangerous side of Nature, which arouses biophobic reactions in us. Reconnecting with Nature does not mean re-turning to the Palaeolithic hunter-gatherer lifestyle but knowing and valuing those aspects that allow us to recover our physical and mental equilibrium more quickly and efficiently. This will be the test bench for Biophilic Design.

Acknowledgements The authors wish to thank Silvia Barbiero for her useful insight into the neurophysiology of biophilia.

References

1. Arranz-Otaegui A, Gonzalez carretero L, Ramsey MN, Fuller DQ, Richter T (2018) Archaeobotanical evidence reveals the origins of bread 14,400 years ago in northeastern Jordan. PNAS. https://doi.org/10.1073/pnas.1801071115
2. Barbiero G (2011) Biophilia and Gaia. Two Hypotheses for an Affective Ecology. J. Biourbanism 1:11–27
3. Barbiero G, Berto R (2016) Introduzione alla Biofilia. Roma, IT: Carocci
4. Barbiero G, Berto R (2018) From biophilia to naturalist intelligence passing through perceived restorativeness and connection to nature. Ann Rev Res 3(1):555604
5. Barbiero G, Berto R, Freire DD, Ferrando M, Camino E (2014) Unveiling biophilia in children using active silence training: an experimental approach. Vis Sustain 1:31–38
6. Barbiero G, Marconato C (2016) Biophilia as emotion. Vis Sustain 6:45–51
7. Beatley T (2011) Biophilic cities: what are they? In: Washington DC (ed) Biophilic Cities. Island Press, pp 45–81
8. Berto R, Barbiero G (2014) Mindful silence produces long lasting attentional performance in children. Vis Sustain 2:49–60
9. Berto R, Barbiero G (2017) How the psychological benefits associated with exposure to Nature can affect pro-environmental behaviour. Ann. Cogn. Sci. 1:16–20
10. Berto R, Barbiero G (2017) The biophilic quality index: a tool to improve a building from "Green" to restorative. Vis Sustain 8:38–45
11. Berto R, Barbiero G, Barbiero P, Senes G (2018) Individual's connection to nature can affect perceived restorativeness of natural environments. Some Observations about Biophilia. Behav Sci 8:34
12. Berto R, Barbiero G, Pasini M, Unema P (2015) Biophilic design triggers fascination and enhances psychological restoration in the urban environment. J Biourbanism 1:26–35
13. Berto R, Pasini M, Barbiero G (2015) How does psychological restoration work in children? An exploratory study. J Child Adolesc Behav 3:1–9
14. Browning WD, Ryan CO, Clancy JO (2014) 14 Patterns of biophilic design. Terrapin Bright Green LLC, New York

15. Browning WD, Ryan CO (2020) Nature inside: a biophilic design guide. RIBA Publishing, London
16. Buss D (2019) Evolutionary psychology: the new science of the mind, 6th edn. Taylor and Frances, Routledge, New York
17. Crutzen PJ (2006) The "Anthropocene." In: Ehlers E, Krafft T (eds) Earth system science in the anthropocene. Heidelberg, Springer, Berlin, pp 13–18
18. Diamond J (1998) Guns, germs and steel: a short history of everybody for the last 13,000 years. Vintage, New York
19. Harvard Medical School (2018) Understanding the stress response. Chronic activation of this survival mechanism impairs health. https://www.health.harvard.edu/staying-healthy/understanding-the-stress-response. Accessed 28 July 2020
20. Kant I (1790) Kritik der Urteilskraft. Berlin und Libau, Verlag Lagarde und Friedrich
21. Kaplan S (1995) The restorative effects of nature: toward an integrative framework. J Env Psy 15:169–182
22. Kellert, S (1997) *Kinship to Mastery. Biophilia in Human Evolution and Development.* Washington, DC, Island Press.
23. Kellert S (2008) Dimensions, elements and attributes of biophilic design. In: Kellert SR, Heerwagen J, Mador Biophilic design, Hoboken, NJ, Wiley, pp 3–19
24. Kellert S (2018) Nature by design. Yale University Press, New Haven
25. Kellert, S. and Heerwagen, J., (2008) *Preface.* In *Biophilic Design,* eds. S.R. Kellert, J. Heerwagen, P., Mador. Hoboken, NJ, John Wiley & Sons, pp. vii-ix.
26. Kellert S, Heerwagen J, Mador P (eds) (2008) Biophilic design: the theory, science, and practice of bringing buildings to life. Wiley, Hoboken, NJ
27. Kellert S, Wilson EO (eds) (1993) The biophilia hypothesis. Island Press, Washington DC
28. Kreibig SD (2010) Autonomic nervous system activity in emotion: a review. Biol Psychol 84(3):394–421
29. Orians GH, Heerwagen JH (1992) Evolved responses to landscapes. In: Barkow JH, Cosmides L, Tooby J (eds) The adapted mind: evolutionary psychology and the generation of culture. Oxford University Press, New York pp 555–579
30. Shimizu H, Okabe M (2007) Evolutionary origin of autonomic regulation of physiological activities in vertebrate phyla. J Comp Physiol A 193:1013–1019
31. Söderlund J (2019) The emergence of biophilic design; Cities and Nature. Springer Nature, Switzerland AG
32. Ulrich, R. (1993) Biophilia, biophobia and natural landscapes. In: Kellert S, Wilson EO The Biophilia hypothesis. Washington DC, Island Press, pp 73–137
33. Wilson EO (1984) Biophilia. MA, Harvard University Press, Cambridge
34. Wilson EO (1993) Biophilia and the conservation ethic. In: Kellert S, Wilson EO (eds) The biophilia hypothesis. Press, Washington DC, Island, pp 31–41
35. Worldbank (2018). https://data.worldbank.org/indicator/SP.URB.TOTL.IN.ZS

Growing the Seeds of Well-Being in the Garden

Alessandra Chermaz

1 Introduction

Planning urban green areas not only brings ecological/environmental advantages, but it also helps preventing and relieving some of the ever-growing cases of psycho-physical diseases that affect our increasingly human-centered communities, and that are becoming a social problem.

Life's efforts and the constant exposure to stressful situations often lead to phases of emotional distress which, in turn, increasingly lead to depression, anxiety and panic. And what's more alarming, is the fact that this growing also affects the younger population, from 10 to 24 years of age. Experts say that by 2020 these illnesses will become endemic, and will turn out to be the second cause of invalidity worldwide.

Working as horticultural therapist with physically and mentally disabled patients (intellectually and mentally disabled, people suffering from Down syndrome, multiple sclerosis, autism, and pluri-disabled blind and visually impaired people) demonstrates how the contact with nature brings various positive effects: refining listening and relational skills; improving group-work abilities; keeping aggressiveness under control; increasing concentration and focus on one's objectives; to relieving tension; etc. The therapy's ultimate goal was not only just to improve some of the individual patient's afflictions, but their quality of life as a whole, so as to favor the person's integration in the local community.

A. Chermaz (✉)
Horticultural Therapist Freelance, Trieste, Italy
e-mail: al.chermaz@gmail.com

2   The Impact of Nature on People

When one observes nature in all its majesty or interacts with it in various activities related to cultivation/reproduction, a deep calm sets in, heartbeats slow down, and one reaches a state of grace called "flow state" in which energy flows perfectly and one is outside everyday reality. It is very similar to a meditative state, where stress is reduced and people become connected with their activity. The person loses track of time, acquires the ability to accept stillness and silence. Meditation techniques are very useful in defining a new landscape the existence, creating a different emotional plane, learning gratitude and appreciating life in its most subtle undertones. Their positive effects are not confined to the moment of practice, but persist in daily life. A study has shown that the positive effects obtained in a few months' practice (e.g. reduced anxiety), can persist for over three year [1].

Most people think it impossible to plan some time each day to relieve the day's stress by inducing a haven of calm, indulging in something they really like. Most people are overloaded and barely get to carry out the chores necessary for their "daily survival". However, if we took a good look at our lives, we would realize that we waste a lot of our precious time. For instance, by indulging on social media, watching TV shows that rather enhance anxiety and fear, hindering our efforts to relax after a day's work. Recent studies show that every 11 min people seek some sort of distraction; check cell phones an average of 85 times a day, and e-mailboxes about every 5 min. It has been calculated that it takes as much as 25 min to get back to a state of maximum concentration. Distraction is often actively sought and tends to become compulsory [9].

In this regard, caring for a garden, or even just a balcony with its flower pots, helps us get out of this state and cut loose from our addiction. Several studies have shown that plants have the power to capture involuntary attention. Their lives often depend from the person and we realize that we cannot leave them to fend for themselves. Plants need person devotion, however they will die. So when people go out on balcony or in garden to water them, they get caught up by every single undertone (all those little changes that nature brings about) and it becomes possible to leave the outside world in a stand-by position. All the stress of the day fades away when we look after our plants. Moreover, we get the chance to train our mind and to investigate and care for details and skills that are constantly under strain in our burner society. Satisfactory results in gardening can only be obtained through perseverance. Therefore, the ability to observe and to persevere are fundamental in order to obtain good outcomes in a garden; the rest doesn't care too much.

2.1   Caring for Plants

Extraordinary performances and successful lives are often ascribed to special qualities or innate gifts. However, many studies on high performance show that the only talent that really makes the difference is the ability to bear an incredibly high number of hours' practice, which, in turn, increasing the ability to bearing physical and mental fatigue. Our brain needs to concentrate for long time spans on a single task, to be able to carry it out with good quality. Researchers have estimated that it takes 10,000 h' deliberating practice in a specific field, in order to reach excellence. Unfortunately, most of the time people stop before they can get to reap the benefits of practice, since they lack patience.

The perseverance needed to obtain good results in caring plants, therefore, help training these abilities marvelously. Another aspect that must not be underestimated is the fact that working with plants trains delayed satisfaction. All people feel the pressure of having to obtain results with all possible haste. Nonetheless, there are times in which people have to work on more complex and lengthy projects, and need to keep a high level of concentration and focus on objectives for longer periods. Without these skills and with no training, is possible finding very hard to handle the situation, leading to great frustration. At the same time, planning garden implies to wait at least five years before it is fully "mature". Continuous adjustments will be made, driven by desire to make improvements. Season after season, the sense of proportion, knowledge of shape, color and foliar texture schemes will be improved. It might often happen that what planting in a certain position doesn't give the results expected at the beginning. The match with the surrounding plants might not be as expected: it might lack balance, heights and proportions may not be as we desired, exposure to light and shade might not be the right one for the best growth, etc. Even though the desire to see how the plants will look like in the new sites chosen for them, before transplanting them, patient is needed to reach their rest period, so as to harm them as little as possible. In the meantime, is it possible to envision them in the new site, imagining our future garden and its new harmonious equilibrium, anticipating our future joy. The patient awaiting for nature's timing is helpful in exercise the imagination and will keep training the ability to bear medium and long-term delays. Neuroscientists have demonstrated that delays can induce the release of "happy hormones" such as dopamine, as much as the positive episode itself [4]. The continuous progresses will raise pleasure and joy, in relation to small, but steady aesthetic improvements. A 'sense of achievement' will be gained, an aspect of well-being that brings about a feeling of progress, mastery, skill and competence. An experience that might be compared to parents' lesson: immediately granting to children all wished denies them the pleasure of delay.

2.2 Plants as a Medicine for Elderly

The lack of expectations for the future, often leads to apathy, asthenia and depression in the elderly, turning them into passive elements of our society. They no longer make decisions, and just endure life, becoming more emotionally fragile. However what makes hard aging, is resignation, the loss of scope and role and the lack of an interior motivation. Growing a small raised vegetable garden, planting brightly-colored flowers, helps in fighting against a deflected mood, bringing an impulsion to go out into the open air and enjoy the spring sun, develops a sense of caring and calls for little, everyday decisions that help keeping the brain active. Moreover, the heed and care that are necessary to grow plants keep seniors away from constantly thinking about the chronic pain that typically affects the last part of our lives. Planting perennials means waiting until the next season before seeing them blooming and luxuriant; the elderly will thus gain a sense of scope, and look forward to the future.

Last but not least, it is fundamental to not underestimate the important physical exercise that caring a vegetable or flower garden implies s, according to the lot's dimensions. Physical exercise promotes the proper presence of neurotransmitters in our brain, which stabilize mood, raise self-esteem and help perceiving ourselves as strong and able.

2.3   Nature as a Therapy for Life

What really makes the difference in obtaining the desired results is not will-power, but simply our joy in undertaking the task itself. If we just focus on the results we will not be able to put up with the efforts they call for. On the other hand, these efforts become easy when we spend time looking after our garden, appreciating every single errand and gaining an immediate sense of fulfillment as we undertake it.

Planning a garden calls for creativity and a sense of beauty. The greatest garden-designers were artists, as Carl Burle Marx, for example. Creative talent is needed to fulfill dreams, having useful ideas, escaping from ordinary dullness and redis-covering the riches we all have inside. Creative-acting brings about a feeling of well-being, almost euphoria. Studies on hemispheric specialization demonstrate that communication between the two hemispheres reaches its peak during the production of ideas. In fact, creative thought favors efficient and fruitful cooperation between the two hemispheres.

Contact with nature in all its majesty, takes back to the childhood and to experience the state of wonder a; a state of mind reached when seeing and discovering something beautiful, a mood rarely feel as adults. Nature as well as gardens are forms of art that strike a number of emotional notes in human beings, and can therefore give way to many different responses. It is impossible not to be touched by such beauty and not be bewitched, this helps to experience the true essence of life.

It is important to nurture amazement since it can turn out to be the answer to many problems, as it helps regaining enthusiasm and vitality. Indeed, lack of vitality is a condition that affects a large part of the population nowadays.

Often live overwhelm our physical and mental powers. It can be called "hamster life", where people are trapped in their running wheel, rushing all day long, and never have the chance to get your head out or stop and think things over. We are all victims of stress, the ultimate killer of vitality and well-being. It slows down neuron production, reduces levels of serotonin and dopamine (mood regulators), and stimulates the amygdala, at the same time reducing the hippocampus function, thus making people absent-minded and exhausted [6–8]. Escaping continuous stress isn't easy, but nature can be of great support. When we take a walk, when we take care of our plants, sow seeds with our children or just sit on a bench observing creation, the whole world stops and our pace of life slows down. We thus get the chance to savor calm, get in touch with ourselves, and let our thoughts run free.

Slowness is certainly one of the greatest victims of our time and our psychological well-being pays the price for it, as shown by increasing number of children under therapy [8]. Children need to live in a slow world, where the leading role is played by leisure, make-believe, easiness, diversions and hobbies; all things that are fundamental for a harmonic growth. Playing with seeds can be of great help in developing children's subtle manual skills. By working with differently sized seeds and learning how to handle very small ones such as linen or Nigella, children expand their dexterity. This, in turn, will improve their handwriting, a skill that suffers the consequences of computer typing.

Nature offers countless fillips; a lifetime isn't enough to learn all its teachings. It stirs our Spirit of enquiry, a very important characteristic if you want to achieve high goals in life. Another important aspect of working with plants is that it training you in following rules. In growing gardens, is important to learn a number of techniques, which ask for a certain order in execution, and put up with the rules set by climate, soil, seasons and other such factors. At the same time, it is also fundamental to learn the correct ways to tackle unexpected problems, such as fungi infection, or infestation by noxious insects. Steadiness in following the necessary steps helps improving the ability in carrying out a task and observing rules, which could turn out to be very helpful in our social and working environments. When designing a garden, an option is to include an area for a family vegetable garden. It takes dedication, commitment and learning. Year after year the expertise grows: refining our skills; learning how to improve the soil, and choosing the vegetables that best suit climate and family needs. Only by growing vegetables is it possible to appreciate their seasonality (nowadays supermarkets offer everything at any time of the year). Therefore, people get accustomed to eating healthy, being aware of what is or isn't good, and controlling our craving for junk food, alcohol and smoke, which are all attempts at escaping for an overloaded mind. Moreover, a closer attention can be developed to the environment, questioning the use of polluting products with all their substantial harm. Pesticides and weed killers ultimately undermine all our efforts to grow a vegetable garden. In the end, even in a small way, all people learning they can all contribute to a better ecosystem.

A vegetable garden has to be carefully designed: it shouldn't be so big as to exceed capabilities, and it must give the right yield for family, otherwise, the fatigue will bury the pleasure of growing it. Its dimensions will have to be reduced when the person grow older, and it should preferably be raised, so as to avoid pointless strain on the backs, legs and arms. High quality results to gratify efforts depend on the accuracy in choosing plants and soil. A raised vegetable garden makes it easier to select the correct substratum. If the soil has to be enriched in an in-ground vegetable garden that has been treated with chemical fertilizers, it would represent a long and hard job. In fact it takes an average of 10 years to improve texture, humus and nutrients needed to be well moisturized and still drain and aerate correctly.

Having a garden implies making compost, which it's quite simple and fulfilling. Making compost isn't a hard job, but certainly at the beginning it requires study, acquire the right skills and following the process until get it. There are all sorts of compost-bins on sale, one of the best has a tumbler inside, by turning it is possible to aerate and at the same time mash its contents, thus accelerating the process. In this way compost is always ready all year long.

3 The Five Senses in Relation to the Nature

Previous experiences of the author, with pluri-disabled blind and visually impaired people, highlighted that senses are not used completely. It's not always true that blindness enhances other senses, it all depends on the person's first years' upbringing. As well know, parents, nowadays tend to be overprotective, are even more if their child suffers from some sort of psycho-physical problem. For example, none of the visually disabled children and teenagers know from the author in her work, had ever sat on the grass or put their bare hands in the soil. Therefore, gardening give them some basic tactile, olfactory and taste knowledge, in order to help them be as self-reliant as possible, and reach the highest quality of life they could.

Among all senses, smell is probably the one that nowadays most suffers from pollution. We hardly perceive smells, as we rarely use this sense. People usually perceiving only the very strong and lasting artificial lab-made odors, that hit the olfactory epithelium cells located in the upper-rear lining of our nasal cavities. Nature's more subtle and less persistent odors, on the contrary, are more difficult to perceive without a proper training.

The center (Regional Institute for the Blind in Trieste) where the author practiced as a therapist had a flower garden, a vegetable garden and a small orchard. The grounds covered an area of 6,000 square meters and were designed for visually disabled users. The area intended for olfactory practice was covered by what the author call "the big five": sage (Salvia officinalis), thyme (Thymus vulgaris), rosemary (Rosmarinus officinalis), lavender (Lavandula officinalis) and oregano (Origanum vulgare). Even though these herbs are well suited to the Mediterranean climate and rich in essential oils, getting acquainted with just five plants is limiting and doesn't contribute to self-reliance. The same narrow design is present in many gardens and paths for the blind around Italy, and this represents how little we know about these people's everyday life and needs. Therefore it was decided to broaden the olfactory range of the garden, adding other herbs we commonly use in our cuisine, such as: mint (Mentha spicata), peppermint (Mentha piperita), lemon balm (Melissa officinalis), lesser calamint (Satureja calamintha), marjoram (Origanum majorana), lemongrass (Cymbopogon citratus), hyssop (Hissopus officinalis), dill (Anethum graveolens), sorrel (Rumex acetosa), anise (Pimpinella anisum), wormwood (Artemisia absinthium), basil (Italian-

Ocimum basilicum, African-Ocimum gratissimum, Greek—Ocimum minimum and mountain b.—Ocimum basilicum 'Alpino'), borage (Borago officinalis), chamomile (Matricaria chamomilla), lemon verbena (Aloysia citrodora), chervil (Anthriscus cerefolium), coriander (Coriandrum sativum), tarragon (Artemisia dracunculus), everlasting herb (Helicrysum italicum), garlic chives (Aethusa cynaium), chives (Allium schoenoprasum), fennel (Foeniculum vulgare), marijoran (Origanum majorana), nepitella (Calamintha nepeta), parsley (Petroselinum crispum), horseradish (Armoracia rusticana), rue (Ruta graveolens), cotton lavender (Santolina chamaecyparissus), savory (Satureja montana), and mountain celery (Levisticum officinale). Blind users practiced a lot distinguishing one from the other, catching the different new scents and recording our new olfactory data. Different recipes was worked out to make the most of these new products, thus learning how to use them in cooking. In this way it was possible to know more about other cultures and the herbs and vegetables they use to flavor their food. For instance, a number of herbs and spices from Tokyo, used in the Japanese cuisine was added to the small farm. Subsequently, the same have been done with the Arab and North African cuisine. Having this wide range of plants at disposal, it was possible to greatly extend olfactory, tactile and taste perception. Then, a laboratory in which to dry our herbs has been set up, used for making different blends for seasoning fish, meat and vegetarian dishes, or for infusions, at times adding also the berries we grew.

Taste is another sense that modern life has impoverished. People have grown accustomed to industrial food, where artificial seasonings, and in particular large amounts of salt and sugar are used specifically to create addiction, and have lost the ability to appreciate delicate flavors. Conversely, at the center, growing a variety of different vegetables in an in-ground and several raised vegetable gardens, accessible to people on wheelchairs, and eating them both raw and cooked helped improving the sense of taste. It is important to pay attention to the different textures of foods, the mouth is in fact an important organ of senses. A freshly picked tomato is crunchy when bitten but also warm and juicy. Eating the ripe fruits just picked from the tree can be electrifying; smelling the fruit, touching it, feeling it in the mouth as you chew it, tasting, it is something unforgettable. Freshly picked vegetables too offer many tactile experiences. They stimulate smell, taste and even hearing. For instance, the noises a firm and crispy freshly picked head of lettuce makes when plucking its leaves to wash it, and when eating it.

Gardening also offers many opportunities to improve touch. Just consider all the raw materials we use for plant reproduction: fine and coarse grained sand, rich compost, bark, vermiculite, perlite, topsoil, dry or pelleted manure, etc. Moreover, it is possible to grow plants with different kinds of leaves: soft, hairy, light or thick and tough, sharp or sticky, trees with different barks, to confer a refined and unusual look to the garden. Plants with peculiar tactile traits can also help the blind, turning into a point of reference for their orientation in space.

There are some features that must be taken into consideration in designing a garden accessible to the blind. One of them is the choice of paving: the lighter ones, as a matter of fact, risk excessive light reverberation. Normally-sighted people may just feel slightly bothered by it, but to a visually impaired person it might be so disturbing as to put him off visiting the garden. Tactile signals are also important, and in-ground vegetable gardens must always have some sort of raised edge, easily detected by blind visitors using a white cane, to avoid accidentally trampling on the beds. Pathways should always be clearly marked and identifiable through plantar tactile sensitivity. They must never be bare earth, and should follow a precise geometrical scheme (parallel to the garden's edges, meeting at right angles); the exit should always be well marked. Raised gardens should never have sharp, dangerous edges; for instance, at the center it took years to round the edges of the raised gardens, since it was not carefully planned.

As well known, shaded areas in a garden are an important aspect. This is even more so in a vegetable garden. Relaxing away from the summer's sun and watching at the work done, for those who still have a visual residuum, adds to the pleasure of the work itself. All you need is a small pergola, where it is possible to grow grapes or early and late kiwi fruit, prolonging the harvest season, thus combining functionality and pleasure.

4 Community Garden

Urban vegetable gardens can turn out to be a good way of enjoying nature when you live in a large or medium-size city. In New York, for instance there are several community gardens: a feature that is just beginning to catch on in Italy, but in the U.S.A. started back in the late 1800s, especially in New York, to address the needs of the lower classes, in times of crisis. The environmental movement and political activism of the '60 s and '70 s brought them new vigor, and by now they are a well-seated custom in the United States. There are long waiting lists to gain access to these shared, self-managed gardens, as they have become very popular. If the successful tenderer neglects caring for his lot for over 6 months, the lot is reassigned to whoever is first on the waiting-list. Community gardens are usually set-up in vacant lots, where a building has been demolished and a new one is pending. Working there gives a feeling of impermanence, since bulldozers might arrive at any time and wipe them out. People are therefore spurred into living in the moment. Community gardens are usually divided into two separate areas: the flower garden where everybody works together on one side, and the vegetable garden where each one has his own lot, on the other. The lots are quite small (about 2 m by 6 m), in order to give more people the chance to have a piece of land in which growing their crop, and also to make it easier for them to care for it. It is an excellent formula for our highly human-centered cities, where houses rarely have gardens, and those that do, are very expensive. It is a wonderful way of creating green lungs for the city, and eating healthy food at the same time. Community gardens are praiseworthy under all respects: they also help social aggregation, bringing people together and making them cooperate, turning our cities into more livable places. In these last few years a growing number of foreigners has shown interest in this experience, finding it a good way of savoring again some of their homeland's typical fruits. Exotic community gardens are spreading: Mexican vegetables, Indian fruits and African tubers are ever more frequent, climate permitting of course. Community gardens, thus, turn out to be a source of food for the body and the soul, a place of moral training, an aid for climate adjustment, a school of ownership and respect, and a way of developing a sense of responsibility. The exchange of agricultural and cooking information helps acceptance and integration of people from different countries, who often risk marginalization.

5 Horticultural Therapy for Companies

Horticultural therapy, can also be used to help the reinstatement of workers who, after long periods of sick leave, lost their role and sense of place within their company. An example of learning process at the R.F.I Company—is described to better understand the different steps. After a period of training, participants have to plan and grow a small garden. They first go through a two weeks' intensive course to learn the basics of agronomy and botany; then they continue to study the fundamental garden design techniques, and only after that they start choosing plants. The experts together with the participants carefully examine the lot in which the garden is to grow. At this point participants take responsibility and decide if it is necessary to improve the structure, the drainage and fertility of its soil by adding humus and sand. They also choose the plants, selecting those that were more compatible with sun exposure, soil, local micro-climate (rainfall, average temperatures, wind, air salinity), soil PH, sizes, appearance, proportions. The aim is to achieving a garden that is interesting and pleasant to look at, possibly all-year-round. Overcoming difficulties in public speaking, each participant can described his/her project to the whole group, justifying his/her choices. In this way everybody can have the chance to learn from the others' thoughts and mistakes. Participants can also join forces and work in group sharing a project. Once the plan was clear, they go on tracing on the ground the plants' positioning, and finally placing them to create a garden. They decide the distribution of tasks for themselves, taking into consideration each one's physical and mental disposition and inclination. Those who have more physical strength took over the more fatiguing tasks, such as planting small trees and bushes and placing down decorative rocks; those who had finer manual skills took care of smaller and more delicate plants. Understanding the importance of an efficient and well-structured organization make it possible to have an equal and efficient distribution of work so that all went smoothly. The result, after this experience is that everybody is manifestly pleased for the work done. Coveting good results, the participants regain their ability to apply themselves in their studies and their will-power, they learn to manage conflict and to mediate, they discover group work and abilities they never thought they could have. The enthusiasm they feel mitigate negative emotions and a new feeling of self-acceptance set in, together with a sense of purpose, understanding and a personal growth. Their self-esteem and their problem-solving abilities are greatly improved, whereas their tendency to judge people (an occupation that tends to constantly engage our minds, wearing out our mental and emotional health) grew weaker. People regain and built up a sense of firm-belonging, and rediscovered the value of solidarity, a value that is often put under strain by competitiveness and all that goes with it.

6 Conclusions

The ever-growing demand for high levels of performance in our modern life, requires a larger number of green areas, accessible to all, where people can mentally and physically restore and recharge.

After a 15 years training horticultural therapists experience, it is possible to note a growing helplessness, not so much in learning the theoretical notions of therapy, as in carrying out the small gestures that are needed for plant growth and reproduction; gestures that often ask for great precision, calm, slowness and patience. Our manual skills, the finer ones in particular, are often quite poor; tasks are carried out hurriedly and lightly, there is no care for details. New generations have grown in a throwaway world and this way of living and thinking has had repercussions in many fields. It won't be easy to recover all the lost knowledge.

People are victims of a phenomenon called "evolutionary discordance". This takes place when the rate at which the environment changes is higher than the rate at which a certain species gets to adapt to the new conditions, giving rise to a discordance between biological needs and environmental reality [5]. Working with plants can mitigate this impression. Nature has its own timing, it's been so for millions of years, and it can help us taking a break away from everyday pressure.

Epigenetics has clearly shown that most of our physical health depends on the way we deal with our mind, our food, our physical activities. The brain adjusts to recurring thoughts and behavior, thanks to a process called neuroplasticity. The more people repeat a certain behavior, the stronger the nerve structure involved in it becomes, just as muscles grow stronger with training. Recent studies highlight that most aspects of genetics are subject to a strong external regulation, epigenetics. In short, the stimuli we give to our DNA influence its activity to the point that any innate aptitude may become insignificant. Growing a flower or vegetable garden helps us establish new habits, it "gently forces" us into healthy practices and habits, in terms of perseverance, diligence, ability in handling fatigue and frustration, awareness of the quality and quantity of the food we eat, physical activity, thus helping us determine the quality of our lives, as epigenetics has shown, and live a long and healthy life.

Those who spend time in caring for a flower or vegetable garden are hard workers who consider it important to obtain results from their work, they are organized, efficient and know how to carry out their task. Industry is strictly linked to determination, which in turn leads to perseverance, which increases the chances of obtaining the desired results.

Finally, this activity favors socialization. People who share the love for plants often develop a sense of solidarity, exchanging knowledge, plants, seeds, grafts, surpluses of vegetables or fruit. Neighborhood relations grow to be easier and more collaborative; discord is rare and people tend to help each other.

The ability of savoring life is an art and, possibly, a discipline. It needs different kinds of sensitivity, commitment, and accuracy. If people allow to modern frenzied life to be overwhelming, it will be very hard, if not impossible, to be fully happy, and when we take stock, the results will put us down. Let's us get away from it, then, and grab our shovel.

References

1. Roi Edizioni (2017) Vivere a Pieno, Filippo Ongaro
2. Marini A, Ediore C, Bussole (2016) Che cosa sono le neuroscienze cognitive
3. Andrea G, Rizzoli BUR (2018)Riconquista il tuo tempo
4. Blazer DG, Hernandez LM (2006) Genes, behavior and the social environment: moving beyond the nature/nurture debate. DC: Washington National Academies Press
5. Begley S, Davidson R, Penguin NY (2012)The emotional life of your brain
6. Stephen L (1995) Scent in your garden. Frances Lincoln Limited
7. Marcus CC, Barnes (1999) Marni Healing gardens. Therapeutic benefits and design recommendations. Wiley
8. Hank B (2009) Gardens for the Senses Winner Enterprises
9. Brendon B, Roi E (2017) Le abitudini per l'alta prestazione
10. Chermaz A (2007) Chi cura chi? La riabilitazione attraverso la natura. EBook -20Les plantes et leur bienfaits. Marie France Michalon. Edition Flammarion, Paris
11. Milano (2007) Giocare con il tatto-per un educazione plurisensoriale secondo il metodo Bruno Munari. Beba Rastrelli, Franco Angeli/Le Comete Edizioni
12. Lemonick M (2005) The biology of joy time
13. John J Miller, Ken Fletcher, Jon Kabat Zinn (1995) Gen hosp psychiatry 17(3):192–200
14. Tafet George E Conference paper 22nd Annual International Conference of IEEE Engineering in medicine and biology
15. Issa G (2010) Neurobiol Dis 39 (3):327–333
16. Ghos S (2013) J Neurosci 33(17):7234–7244
17. L'Adige.it (newspaper)

Design of Natural Places for Care: Strategies and Case Studies

Monica Botta

1 Introduction: Gardens for Health

The role of the healing gardens as a proper therapeutic tool within the care context has achieved a particular relevance at the international level during the last decades. Furthermore, the positioning and typology of the various gardens and the treatments undertaken may be different depending on cultural and economic issues concerning the management of health-care institutions.

The design of therapeutic gardens currently relying on Evidence Based Designed (EBD) methodologies, demonstrating that green areas, as a variable of the built environment, can influence on various health-related outcomes [1]. In particular, the main findings are related to the work of Clare Cooper Marcus [2, 3]; and defined by the AHTA—American Horticultural Therapy Association [4] by means guidelines in force for more than twenty years. However, the spread of therapeutic gardens is characterised by a different and heterogeneous contextualisation and in the Italian context, it is still under development.

1.1 *Italian Context*

In Italy, the realisation, the use and the real benefits of therapeutic gardens is related to the social and healthcare sector more than to the hospital context. There are virtual examples of therapeutic gardens within hospitals, but the system of use, care and management of such spaces is very complex. In the last decade, the inclusion of these gardens in nursing homes, hospices, day-care facilities for people with disabilities and elderly people, Alzheimer centres etc. is increasingly becoming an added value and a

M. Botta (✉)
Landscape Architect, Bellinzago Novarese (NO), Italy
e-mail: m.botta@monicabotta.com

real extension of the therapy. The privatisation and leaner management within these structures allow to managing, monitoring and assessing the therapeutic efficiency of the gardens.

The need to find a method in Italy

Understanding the features of the therapeutic gardens and the treatment goals is a crucial point. The healthcare rehabilitative treatment of people is changing direction towards a holistic approach. Scientific research studies upon the subject, in-depth analysis and realisations contribute to confirm and improve the evidence concerning the use of green areas and nature as means for the well-being.

The introduction, within the Italian context, of project standards for therapeutic gardens such as those used by the AHTA [4], in addition to a medical monitoring properly set up and validated, may give credibility to methodologies already consolidated within the international context. In this regard, studies aimed to measure the impact of the built environment on people health, through qualitative and quantitative indicators, are growing [5–8].

Moreover, there is a need to integrate in the current legislation, the project guidelines and standards with specific indications about this kind of green areas a. In fact, often designers do not take into consideration the nature, the gardens as social elements of aggregation and of well-being.

1.2 Gardens for People with Dementia

Gardens for people with dementia and Alzheimer aim to: reduce the temporal and spatial disorientation; contain the "wandering"; stimulate the residual capacity; compensate the cognitive deficits and functional caused by the dementia; oppose restlessness, irritability and aggressiveness; abolish the appeal to the restraining; reduce the use of medicines for the treatment of apathy, sadness, depression; maintain the levels of autonomy; increase the individual self-esteem; improve the tone of the humour; improve of the state physical and mental and encourage the socialisation.

In particular, there is a series of specific design elements of gardens for people with dementia, that are fundamental in order to make them healthy places. Specifically, the key points related to the realisation of these spaces can be described by the following aspects: *Sustainability; Orientation; Accessibility; Socialisation; Meaningful activity; Reminiscent (History); Sensory stimulation; Security/Comfort.*

In the Italian context, these indications are integrated with the concept of the "Genius loci", which in the design field, means that the strong territorial, landscape and historic dimension implies a design response calibrated on the basis of the place where the gardens will be included. From the North to the South, Italy has an incredible variety of vegetal species, architectural typologies (from materials to construction methods) and social fabrics, which are the emblem of the locations. For this reason, the inclusion of a therapeutic garden able to satisfy the different forms of comfort is a fundamental action.

Therapeutic programme in the gardens

A well designed garden and attentive to the needs of the ill person, becomes the ideal place for supporting the treatment and therapeutic programmes of sensorial and cognitive stimulation. It can be the place where people alleviate the stress, unleash the wandering, and find the way to activate curiosity towards natural elements. Both in a day-care centre and in an accommodation facility, the garden can be enjoyed during the day and often also in the evening and night hours.

Different activities and therapies can be performed within the green context such as *Pet therapy; Horticultural Therapy; Gardening; Physiotherapy; Readings; Laboratory; Praying*, etc. Furthermore, a monitoring and a therapy path should be developed in relation to the use of the garden itself.

International research suggests that the physical access and the view of nature help overall people to:

- recovering more rapidly from illness conditions;
- reducing the stress and lower blood pressure;
- maintaining the circadian rhythms (sleep cycle/wake cycle);
- the natural absorption of D vitamin (considering that the exposure to sun light for brief periods of time is important to maintain strong bones).

2   Case Study: Healing Garden—Garden of Happiness

2.1   Design Methodology

The project of the garden takes into account from the initial design phases the needs of the final users to defining the project layout. In line with Design for All strategy, which is the design for human diversity, inclusion and equality, different users were involved during the design process, through a dialogue at every stage of the design process, in order to understand their needs [9]. In this regard, preliminary investigations, such as focus groups, interviews, study of project guidelines and of the territory have been addressed…

At first, a guests' **background analysis**, through documentation provided by the administration of the structure (number of people, age, sex, pathologies, autonomy) have been carried out together with a **survey on the activities performed** within the social and healthcare unit (laboratories, meetings, non-pharmacological therapies, physiotherapy) (Table 1 and Fig. 1).

Table 1 Garden data

Healing Garden: Garden of Happiness—Ferrara	
Name	Garden of Happiness
Site	Ferrara—Italy
Designer	Monica Botta landscape architect
Client	Residence Città di Ferrara s.r.l
Realisation	Summer/Autumn 2017
Size	2,500 m^2
Users	n. 180 elderly with dementia
Health workers	n. 140
Relatives	n. 250/350

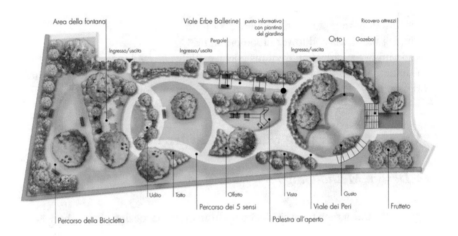

Fig. 1 Plan of the garden

In particular, three different categories of focus groups were performed.

- **Two focus groups with the administrative staff** provide indications about: the aims of the therapeutic garden; the inclusion of different therapeutic spaces following the examination of the draft project and the definition of furniture and colours.
- **Three focus groups with five physiotherapists** provide useful insights on: the definition of the space for the gym; the types of movements and exercises feasible outdoor and the modifications to the rehabilitation equipment included.
- **One focus group with 20 guests** of the structure needed for: selecting the vegetation; discovering ideas for the use of the garden; verifying the areas envisaged by the designer and choosing of the garden's name.

In addition to the focus groups, a **visual observation** of people's behaviour and of the use of the spaces was carried out within the social and healthcare structure.

During the analysis, **sample survey questions** were asked to different guests in order to better understand their personal needs [9]. Moreover, during the realisation, the guests were constantly updated on the progress of the works in order to better involve them in the future use of the garden.

Finally, the analysis of the city of Ferrara features, based on slow mobility and the use of bicycles [7], brings to the integration of a path which follows this type of transport and the local culture.

2.2 Garden Design Description

With an area of about 2,500 mq, the Garden of Happiness is composed of a series circular paths that run around playgrounds dedicated to specific activities. It is accessible from pedestrian entrances and is located within a protected area (Table 2).

A **path with draining pavement** leads to waiting areas used as spaces for physical activity. Along the path, there is the **Street of Dancing Grass** (Fig. 2), including two small pergolas with seats, is adorned by the movement of grasses and perennial herbs in the wind, stimulating the touch and sight.

The **Vegetable Garden** and **Orchard area** is composed by a circular space with cantilevered horticultural desks and flowerpots, equipped with a mechanised pergola tent, furnished with a large table and chairs.

At the centre of the garden is located an **Outdoor Gym** (Fig. 3), aimed at giving the opportunity for physical exercise, autonomously or with the help of physiotherapists. A draft project originally included a boules alley instead of the gym, but the final project was changed for different reasons, such as: only 20 guests out of 180 are autonomous (not in a wheelchair); guests can throw only plastic balls up to a maximum distance of 3–4 m and the number of women is higher than the number of men, which use 3 to play this game.

Table 2 Garden areas and dimensions

Areas of the garden	Dimensions	Activity
Vegetable garden and gazebo	50 mq	Horticultural therapy, gardening, meeting, laboratory
Orchard	30 mq	Horticultural therapy
Five senses path	40 m	Sensorial stimulation, gardening
Outdoor fitness	30 mq	Rehabilitative activity in autonomy and with the aid of the physiotherapists
Fountain area	20 mq	Socialisation, laboratory, meeting
Street of the dancing grass	20 m	Stimulation of the touch
Way of the bicycle	40 m	Motor activity, rehabilitative, socialization, stimulus to the memory of the place

Fig. 2 Street of the dancing grass

Fig. 3 Outdoor gym

In the southern part, at the **Five Senses Path** stimulating vegetation is contained within almond-shaped flowerbeds, such as the taste flowerbed (containing small fruits); the sight flowerbed (with roses blooming until November); the smell flowerbed (with aromatic herbs); the touch flowerbed (with vegetation suited for being touched) and the space of hearing with a fountain.

The **Fountain**, an architectural element with 12 gushes, ad-hoc designed, defines a polyvalent space for spiritual gatherings, meetings, assemblies, equipped with tables and chairs.

Fig. 4 Way of the bicycle

Behind this area, the **Way of the Bicycle** (Fig. 4) is made of anti-trauma rubber, which refers to the city of Ferrara, where this mean of transportation is widely used. Some sculptures—designed by architect Botta—are located along the path, with handlebars and bells allowing physical training.

Particular care was devoted to support and safety elements within the garden adopted to facilitate the usability: certified draining and anti-slip pavements, anti-trauma-draining pavements, vertical signs, handrails, emergency buttons and evening/night lighting. Furnishing elements in painted galvanised steel, such as single seats, benches and tables, have been introduced. They are non-fixed, movable and identifiable in colour as the two blue light fountains.

All the vegetation included aims to introduce elements of shadow, sensorial stimulations, visual points of positive impact and at creating a constant transformation and enjoyment of the garden in all seasons, as well as attracting butterflies and small insects in order to foster life in the natural space.

2.3 Results: Therapeutic Target

The targets of this healing garden, established with the client, are diversified and aim at enhancing the quality of life and work of the guests and staff:

- providing a therapeutic environment;
- increasing the places of socialisation;
- performing different activities (e.g. horticultural therapy, gardening, reading, Physiotherapy, etc.) increasing motor and rehabilitative activities (outdoor fitness);
- providing deviation from dementia or illness symptoms;
- allowing the wandering;
- finding quiet and privacy;

- increasing ludic and recreational activities related to nature (gardening horticulture);
- providing sensorial stimulation.

The following **benefits to senior** users were deduced through interviews with the staff and observation: mood improvement; decision to spend half or full day in the garden; stimulation of curiosity and observation; complete autonomy in the use of the spaces of the garden; improvement of psycho-physical recovery and lowering of the tensions among the guests.

In addition, the **benefits to workers** and **relatives** of the garden can be summed up with the following characteristics: discharge of working tensions by the staff; break and relax place for the staff; leisure place for the relatives and extension of the pause with walking of relatives and guests within the garden.

3 Case Study: Alzheimer Garden "Il Faggio"

3.1 Design Methodology

The project of the garden takes into account "Evidence Based Data" guidelines of the Alzheimer garden in order to define the layout, since the structure was a new construction (Table 3 and Fig. 5).

A brief **preliminary investigation** allows to understand that the Onlus in charge of the management of the day-care centre was not aware of the therapies that could have been undertaken through the use of the garden.

A **briefing** with the Company 'Generali Arredamenti', was fundamental for the introduction of non-pharmacological therapies, such as the Train Therapy, the Multi-sensorial Room and the horticultural desks within the centre. In addition, the

Table 3 Garden data

Alzheimer Garden "Il Faggio"—Salerano Canavese (TO)	
Name	Il Faggio
Site	Salerano Canavese (TO)—Italy
Architectural design	aMDL Michele De Lucchi architect
Garden designer	Monica Botta landscape architect
Client	CasaInsieme onlus
Realisation	Spring 2018
Size	1000 m^2
Users	n. 20 seniors with Alzheimer
Health workers	n. 5
Relatives	n. 30/40

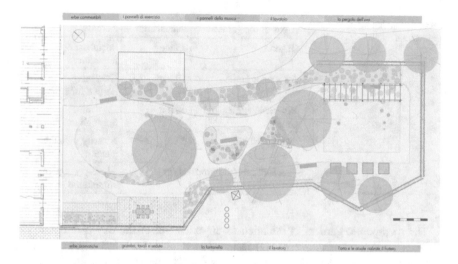

Fig. 5 Plan of the garden

need to integrate the internal therapy programme, introducing different elements complementing the internal proposals, emerged.

The **analysis of the existential background** of the guests was not possible since the structure was new and there were no information about the perspective guests. Therefore, one **focus group with the administrative staff** was performed highlighted the following indications:

- the request of visibility in the overall garden (low vegetation);
- the introduction of wooden furnishing elements (in line with the project features of the centre);
- the request of designing ring paths;
- the need to delimit all the area;
- the issues concerning some pre-existing elements (sewers of the heating system to be covered, presence of significant dimensions, presence of an historical vine pergola).

3.2 Garden Design Description

The pre-existing area where the garden is created is part of the larger park adjoining Villa Sclopis. The area was a green lawn with some trees among which a Douglas fir at the centre, a stone vine pergola along the perimeter, some fruit trees among which a nut tree and some kaki trees (Table 4).

Table 4 Garden areas and
dimensions

Areas of the garden	Dimensions	Activity
Gazebo	25 mq	Socialisation, laboratory
Aromatics herbs	10 mq	Gardening, horticultural therapy
Vegetable garden and pergola	60 mq	Gardening, horticultural therapy
Stone washtub	5 mq	Activity with the water
Panels of rehabilitation	30 m	Sense-perceptive stimulation

The therapeutic garden, is intended as an extension of the day-care centre "la Baracca", in Salerano Canavese (TO) designed by Michele De Lucchi, was conceived by the landscape architect Monica Botta, specialised in the design of healing gardens.

From the porch of the day-care centre, it is possible to see at a glance all the garden and monitor all the guests. For this reason, the garden "Il Faggio", covering an area of more than 1,000 mq, was designed with low-scale vegetation in the central parts of the green area and higher vegetation, coloured and evergreen along the borders.

At the centre of the garden, a large pre-existing fir shadows a large oval space, in part covered by evergreens and flowers and for the remaining part dedicated to the free leisure of the guests as a green lawn in front of the building.

The garden, with a series of ring paths in natural eco-friendly material, contains rest and refreshment points such as the tensile structure with a large table and wooden chairs allowing the guests to rest and perform activities. This space can be reached from the porch through a path running along an area with **aromatic herbs** and edible plants, located near the kitchen in order to be used and manipulated.

Generali Arredamenti's **panels**, for physical exercise and with sound effects, run along the path opposite to the tensile structure, in order to give the opportunity to find different therapeutic spaces also in the garden. Two central green areas hosts large wooden benches.

The entire building is in wood as part of the fence, which studied to give material continuity to the green space. In order to stimulate the memory and remembrance, two water points were included within the garden: an ancient **fountain** and a Valdotain **stone washtub**, aimed at guarantee both the access to the water, and to recall ancient gestures performed by women at typical mountain washtubs.

The garden is closed by a handmade **stone pergola** creating a shadowed path with the vine. This area—with fruit trees, nut tree, plum tree and grapes—is intended to be used as vegetable garden and the green adjoining the pergola will be transformed in a flower meadow during the next autumn. In this area, raised and movable horticultural desks will be located in order to allow horticultural and gardening activities.

The therapeutic garden is aimed at stimulating, through the insertion of grasses and perennial herbs, touch, sight and hearing. Vegetation masses will swing and change colours and consistency during the cycle of seasons. Perfumes and stimulations dedicated to the touch will be especially provided by the aromatic herbs flowerbed. The area of the fountain will also be an incentive for the arrival of butterflies, since different types of plants attracting insects based on their colouring and blooming. All the green area has therefore therapeutic connotations as long as sensorial, motor, evocative, historical and local stimulations support and help the care of people (Figs. 6, 7, and 8).

Fig. 6 Stone washtub and pergola

Fig. 7 Bench

Fig. 8 Panels of
rehabilitation

3.3   Findings: Therapeutic Target

The garden aims to becoming an extension on the treatment place within the day-care centre. In particular, the targets to be pursued through its use by the guests affected by Alzheimer's disease are:

- providing a therapeutic environment;
- providing deviation from the symptoms;
- allowing the wandering;
- socialising or withdrawing;
- finding quiet and privacy;
- increasing physical activities;
- increasing ludic and recreational activities related to nature (gardening horticulture), and according to accessibility for all the different users' needs [9]

The following **benefits to senior users** were deduced through interviews with the staff and observation: mood improvement; allowing the wandering; deciding to spend half or full day in the garden; use of the benches to enjoy nature also during naps; stimulation of curiosity and observation; complete autonomy in the use of the spaces of the garden [10].

Finally, the **benefits to workers** and **relatives** of the garden can be summed up with the following characteristics: reassurance of the staff; enabling to monitor the guests of the centre; discharge of working tensions by the staff; leisure for the relatives without the presence of the related guests; extension of the break with walking of relatives and guests within the garden.

Conclusions The two case studies represent a first attempt to assess the benefits generated by an healing garden on people behaviour [11–15]. From a critical perspective, an Evidence-based survey would be necessary in order to determine the prosthetic effectiveness of the two therapeutic gardens, since observation and interviews took limited time (about 1 year for the first case study and 3 months for the latter). Furthermore, the assessment of the benefits should be performed evaluating the outcome for the guests [16], for the staff and for the relatives. In particular, the assessment should be complemented with evaluations performed by clinicians, with specific attention to the physical and mental benefit resulting from the use of the garden.

Therefore, only through systems of cross analysis between design features and health outcomes [1, 5–7], it is possible to identify an increasing therapeutic value of gardens, quantifying and qualifying the well-being of those who have used and enjoyed them.

References

1. Brambilla A, Capolongo S (2019) Healthy and sustainable hospital evaluation—a review of POE tools for hospital assessment in an evidence-based design framework. Buildings 9(4):76. https://doi.org/10.3390/buildings9040076
2. Marcus CC, Barnes M (1999) Healing gardens. Therapeutic benefits and design recommendations. Wiley, New York
3. Marcus CC, Sachs N (2014) Therapeutic landscapes. An evidence-based approach to designing healing gardens and restorative outdoor spaces. Wiley, New York
4. American Horticultural Therapy Association (1995) Therapeutic gardens characteristics. http://ahta.org/sites/default/files/attached_documents/TherapeuticGardenChracteristic_0. pdf. Accessed 12 Dec 2012
5. Buffoli M, Rebecchi A, Gola M, Favotto A, Procopio GP, Capolongo S (2018) Green soap. A calculation model for improving outdoor air quality in urban contexts and evaluating the benefits to the population's health status. In: Mondini G, Fattinnanzi E, Oppio A, Bottero M, Stanghellini S (eds) Integrated evaluation for the management of contemporary cities. Green energy and technology. Springer, pp 453–467. https://doi.org/10.1007/978-3-319-78271-3_36
6. Mosca EI, Capolongo S (2018) Towards a universal design evaluation for assessing the performance of the built environment. In: Craddock G, Doran C, McNutt L, Rice D (eds) Transforming our world through design, diversity and education: proceedings of universal design and higher education in transformation congress 2018. Studies in health technology and informatics, vol 256, pp 771–779. https://doi.org/10.3233/978-1-61499-923-2-771
7. Rebecchi A, Boati L, Oppio A, Buffoli M, Capolongo S (2016) Measuring the expected increase in cycling in the city of Milan and evaluating the positive effects on the population's health status: a community-based urban planning experience. Ann Ig 28(6):381–391. https://doi.org/10.7416/ai.2016.2120
8. Rebecchi A, Buffoli M, Dettori M, Appolloni L, Azara A, Castiglia P, D'Alessandro D, Capolongo S (2019) Walkable environments and healthy urban moves: urban context features assessment framework experienced in Milan. Sustainability (Switzerland) 11(10):2778. https://doi.org/10.3390/su11102778
9. Mosca EI, Herssens J, Rebecchi A, Capolongo S (2019) Inspiring architects in the application of design for all: knowledge transfer methods and tools. J Access Des All 9(1):1–24. https://doi.org/10.17411/jacces.v9i1.147
10. Boatti L (2017) Healing gardens e horticultural therapy. L'attuale scenario Nazionale e Internazionale. Tesi di laurea Master Universitario di II livello Congiunto in Pianificazione, Programmazione e Progettazione dei Sistemi Ospedalieri e Socio-sanitati – VIII edizione, a.a. 2016/2017
11. Botta M (2017) Giardini protesici per le persone affette da Alzheimer. Med Integr 4:46–50
12. Botta M (2017) La vegetazione e la sua componente terapeutica. Med Integr 6:52–56

13. Botta M (2012) Verde terapeutico. Quei progetti mirati che aiutano a guarire. Assist Anziani 21–27
14. Botta M (2012) Healing garden. Giardino terapeutico per anziani, disabili, bambini. Percorso storico sensoriale, terrazza verde, Orto dei Semplici, percorso fisioterapico. E-volution, Torino
15. Botta M (2018) Caro giardino, prenditi cura di me. Delicate storie di benessere nella natura. LDN editore, Milano
16. Borghi C (2007) Il giardino che cura: il contatto con la natura per ritrovare la salute e migliorare la qualità della vita. Giunti editore, Firenze

Therapeutic Architecture. Assessment Tools and Design Strategies for Healing Gardens Implementation

Andrea Rebecchi, Andrea Brambilla, Monica Botta, Angela Casino, Sara Basta, and Stefano Capolongo

1 Introduction

The World Health Organization (WHO) defined the concept of health as "*a complete state of physical, mental and social wellbeing and not merely the absence of disease*" [1]. Health is not anymore intended as a "*state*" but as a dynamic and multidimensional condition of equilibrium, also related with the capability of each subject of positively interacting with the surrounding environment [2, 3]. In the field of healthcare and socio-sanitary facilities design and research, a consolidated strategy is to enrich hospital spaces through the implementation of soft qualities and therapeutic design elements. Among the different interventions a popular one is the introduction of green open spaces specifically design to address the psycho-physical and psycho-emotional needs of the users, called Therapeutic or Healing Gardens [4–6].

A. Rebecchi (✉) · A. Brambilla · S. Capolongo
Department of Architecture, Built environment and Construction engineering (ABC), Politecnico di Milano, Milan, Italy
e-mail: andrea.rebecchi@polimi.it

A. Brambilla
e-mail: andrea1.brambilla@polimi.it

S. Capolongo
e-mail: stefano.capolongo@polimi.it

M. Botta
Monica Botta Landscape Architect, Bellinzago Novarese (NO), Italy
e-mail: m.botta@monicabotta.com

A. Casino · S. Basta
School of Architecture, urban planning, construction engineering (AUIC), Politecnico di Milano, Milan, Italy
e-mail: angela.casino@mail.polimi.it

S. Basta
e-mail: sara.basta@mail.polimi.it

Those interventions have been mainly considered as places able to positively benefit patients and relatives [7, 8]. As highlighted in several Evidence Based Design (EBD) studies, also medical staff can benefit from the built environment characteristics in several outcomes such as work satisfaction improvement, stress reduction and burnout syndrome challenge [9–11]. Additionally, they spend most of their working days into the healthcare facilities and are the users that might benefit from those intervention for a longer period. Nevertheless, to the best of our knowledge, a limited number of researches have been conducted on the topic.

Therefore, the aim of the research is to investigate the impact that Healing Garden have on doctors and nurses perceived wellbeing and to provide tools and strategies for their design implementations.

In the first part the therapeutic effects of nature and green spaces have been highlighted; then the methodology adopted for data collection and analysis have been described; finally, the results have been discussed and design strategies have been proposed.

2 Theoretical Background

Several theories and researches demonstrated how nature and green space have positive impacts on individual and population health highlighting the role of therapeutic architecture since ancient times. Recent seminal studies by recognized experts from the environmental psychology discipline, defined two major theoretical framework that are still debated today: Ulrich's Theory of Stress Reduction (TRS) and Kaplan brothers' Theory of Attention Regeneration (TRA), both applicable in natural spaces such as healing gardens [12–14]. Specific studies in this direction also highlighted different streams of research focusing on the effect that the nature has on physical symptoms alleviation and also on the capacity to reduce stress and improve comfort values [15–17].

Therapeutic landscapes and healthcare healing environments in general can improve sense of personal control, social support, positive distraction, reduce environmental stressors, connect patients with nature and stimulates feelings [18]. The terms related to therapeutic architecture are indeed widely used and can be generally considered as tools for wellbeing improvement through the built environment. Therefore it is important to consider all the different users: patients, families and especially medical staff and nurses whom wellbeing and satisfaction considerably impact on the quality of medical care and assistance.

3 Purpose and Method

Starting from those considerations and the emerging trends in healthcare architecture and EBD field, the research question is: *Which are the benefits/impacts in terms of self-reported wellbeing that Healing Gardens have on medical staff and nurses?*

In order to address the aforementioned research question, a qualitative empirical study has been conducted on a sample of seven case study selected among national and international examples of Therapeutic Gardens inside socio-sanitary facilities or nursing homes.

In the international scenario, a literature research highlighted the existence of relevant experiences in British and American context about Therapeutic Gardens assessment from the medical staff perspectives. Therefore four case study have been selected:

1. Family Life Centre, Grand Rapids, Michigan (USA)
2. Homewood Health Centre, Guelph, Ontario (USA)
3. Lodge at Broadmead, Victoria, British Columbia (USA)
4. Joseph Weld Hospice, Dorset shire, Dorchester (UK)

Additionally, in the national panorama three specific cases of Therapeutic Gardens inside socio-sanitary facilities recently completed and currently in use have been chosen:

1. Il Girasole, Chiavenna (SO)
2. Punto Service, Bellinzago Novarese (NO)
3. Residence Service, Ferrara (FE)

Each case has been analysed through a set of parameters including: construction year, designer, geometrical dimensions, number of beds, number of workers, architectural and functional characteristics, presence of socialization and rehabilitation spaces, circular paths, boundaries and fences, tables and seating areas, flowers and botanic species and water presence, as suggested in Cooper Marcus work [19].

Additionally, in the Italian example direct surveys, semi-structured interviews, photographic analysis, functional and typological investigation have been useful in this data collection phase (Fig. 1).

Starting from the literature and case study analysis, an assessment tool in the form of a questionnaire has been defined and submitted to the medical staff and nurses of the selected case studies. The instrument is based on the review of several existing surveys available on the aforementioned international case studies and relevant practice around the world. The concept of evaluation in the built environment research is indeed very important and useful especially in the form of Post Occupancy Evaluations (POE), Health Impact Assessments (HIA) and multidimensional analysis tools [20–26]. The questionnaire is composed by 5 closed-ended demographic questions, 7 questions with multiple answers specifically related to the use of healing garden and the self-reported benefits in terms of wellbeing and satisfaction, and 1 final open-ended question. Published results related to interviews conducted in the four international case studies by other scholars have been taken as secondary data.

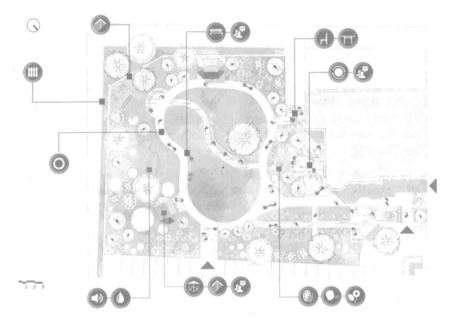

Fig. 1 Example of a case study analysis—Family Life Centre, Grand Rapids, Michigan (USA) (re-design of the case study plans by the Authors)

On the contrary, the direct investigation of national cases allowed to collect primary data from the medical doctors and nurses working in the three Italian facilities. The total sample was composed by 1560 users for the international case and 181 in the national one. In the former the percentage of respondents was 29% while in the latter 23%.

The two sets of data have been elaborated through a comparison matrix in order to achieve a broader view on the topic highlighting the presence or absence of specific therapeutic elements or wellbeing outcomes.

4   Findings and Discussion

Data have been collected and elaborated through a comparison between the international (secondary data) and the national cases (primary data). The most interesting ones are presented and discussed below with specific reference to four domains: daily time usage, purpose of usage, self-reported benefits and overall satisfaction.

Fig. 2 Frequency of usage of the Therapeutic Garden in national and international case studies (authors' elaboration). The diagram are in italian cause they comes from the original survey: the left diagram describe the time spent into the Healing Garden during the week; the right diagram show the safe frequency during the month

4.1   Frequency of Use

The Therapeutic Garden is generally used by the staff less frequently than expected. Indeed, only 30% of the respondents in international cases declared to use the garden on a daily base, while 64% less regularly (weekly, monthly or annually). 7% of the respondents never used it. Similar figures are reported by the national respondents: 31% use the garden on a daily basis (more than 5 or more than 10 times a week), 53% use it less than 5 times per week, and 16% never used it (Fig. 2).

4.2   Daily Time Usage

More than half of the respondents (58%) at the national level declared to use the Therapeutic Garden in the morning between 9:00 and 12:00. This is consistent with the fact that Therapeutic Gardens in the selected case studies are closely linked with the socio-sanitary facilities both visually and physically; therefore, medical staff is more prone to use the open space before the beginning of the working day or during the morning activities.

On the contrary, in Italian scenario only 13% use this space between 12:00 and 14:00 while in international one the figure rise to 36%. A deeper look at the two typologies highlights the presence of shadow areas, tables and seats in the international case studies which might impact the easiness for staff to spend their lunchtime in open air (Fig. 3).

Fig. 3 Daily time usage of
the Therapeutic Garden in
national and international
case studies (authors'
elaboration). The diagram
is in italian cause they comes
from the original survey: the
difference between blue and
light blue is, respectively,
international and national
context

4.3 Purpose of Usage

Most of the respondents (84%) in the Italian scenario use the Therapeutic Garden for
strolling thanks to the presence of path, sensorial walks, aromatic plants, flowerbeds
and water streams. Half of the respondents (52%) use the garden as a place where
to conduct therapeutic activities with the patients. This can happen thanks to several
designed elements and instruments for patients rehabilitation and physical exercises spread around the garden. Additionally 39% stated that use the garden also to
socialize, exploiting the breaks from working activities in open air and some uses
the garden for smoking (between 5 and 10% in both cases). A relevant issue both for
national and international examples is that less than 5% of the respondents use the
Therapeutic Garden for having lunch. Although the data is probably mainly linked
to the working shift organization, also the layout and the lack of appropriate seats
and tables might have impact on that (Fig. 4).

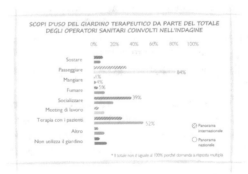

Fig. 4 Purpose of Therapeutic Garden usage in national (full-line) and international (dashed-line)
case studies (authors' elaboration). The diagram is in italian cause they comes from the original
survey: the analyzed alternatives, top-bottom, are staying in the place, walking, eating, smoking,
socializing, lead a business meeting, therapy with patients, other situation or he/she doesn't use the
green space

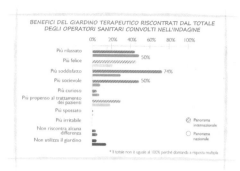

Fig. 5 Self-reported health and wellbeing benefits in national (full-line) and international (dashed-line) case studies (authors' elaboration). The diagram is in italian cause they comes from the original survey: the analyzed alternatives, top-bottom, are more relaxed, more happy, more satisfied, more sociable, more curious, more inclined to the patients' treats, more exhausted, more irritable, he/she finds no difference and he/she doesn't use the green space

4.4 Self-Reported Health and Wellbeing Outcomes

All the respondents that use the Therapeutic Gardens reported an improvement in all the health and wellbeing parameters available in the survey with extremely low percentage of respondents feeling without differences or more tired, and no one feeling more irritable after the use of those spaces. In particular, a relatively high percentage of sanitary staff both in national (50%) and in international setting (45%) reported a more relaxed status after the use of a Therapeutic Garden. Additionally, in USA and UK settings 74% of the medical staff declared to be more satisfied, 50% more friendly and 30% more prone to help the patients. Several design elements such as sensorial greenery, seating areas and flowers are able to improve the quality of the therapeutic gardens and match with the actual expectations of the people working there. Additionally, as reported by the 86% of the national and 91% of the international respondents, the need of looking at the garden from inside emerges as a very important topic for the medical staff and nurses perspectives (Fig. 5).

4.5 Design Strategies

The data collection, analysis, elaboration, discussion and the comparison of the empirical data with the physical characteristics of the different Healing Gardens, allowed to highlight the crucial role of therapeutic green areas in socio-sanitary facilities and nursing homes. This study proposes specific design recommendations and guidelines to improve even more the qualities of those natural spaces.

Starting from the secondary and primary data collected and analyzed along with the technical features highlighted in the case studies, it is possible to frame specific design strategies that can support the quality improvement and design implementation of such nodal spaces with specific regards to 5 macro categories:

1. *Spatial individuation*: it is important to properly define spaces according to the needs of each disease stage, with specific attention to the indoor-outdoor contact and transition, clearly providing socialization and rehabilitation spaces.
2. *Spatial qualities*: Therapeutic Gardens should guarantee a high degree of privacy for the users, as well as safety, especially for the most fragile patients. Space clarity and familiarity, which are fundamental in order to improve the usability of the space, could be achieved through the use of different colors and textures.
3. *Accesses and paths*: the gardens should be accessible in a clear and preferably unique way, with specific regards to disable and fragile users according to Universal Design and Design for All strategies [27]. Paths should be circular or with soft angles and boundaries must be appropriately fenced.
4. *Natural elements*: natural features are obviously fundamental especially with regards to horticultural therapies and gardening. Particular attention must be payed to avoid the use of toxic plants and promoting the implementation of aromatic plants (i.e. lavender, rosemary, wisteria, thyme, lemon balm, sage), safe water elements and natural sounds [28].
5. *Furniture elements*: ornamental and functional equipment should be designed properly to improve orientation and wayfinding and stimulate memory [29]. Shadow areas, tables, fixed and movable sittings, handrails and artificial lights should be regularly implemented.

5 Conclusions

Natural space is very important in the design of healthcare facilities and is widely recognized as a way to improve human health and wellbeing. Evidence Based Design studies consistently highlight the impact that natural solutions and green areas [30] have on individual and population health outcomes. Within the socio-sanitary facilities a crucial role is played by Healing Gardens that can assume the character of a therapeutic architecture within the whole system of spaces. Patients benefits are very important but also medical staff and nurses can improve their wellbeing thanks to the presence and the correct design of those spaces.

The empirical study conducted through the exploitation of secondary data in international case study and the collection of primary data in national case studies, supported these assumptions and enriched the existing body of knowledge on the topics [31].

The research demonstrates that Therapeutic Gardens can be beneficial also for hospital staff which regularly use this space for social and therapeutic activities.

In particular they have a significant impact on perceived satisfaction, relax improvement and willingness to work with the patients.

Future research on the topic are encouraged in order to involve a wider sample of socio-sanitary facilities and nursing homes' Therapeutic Gardens. The investigation could be also implemented in other healthcare facilities such as clinics, territorial centers and general hospitals worldwide.

References

1. Grad FP (2002) The preamble of the Constitution of the World Health Organization: public health classics. Bull World Health Organ: Int J Public Health 80(12):981–984. https://apps.who.int/iris/handle/10665/71722. Accessed 13 Dec 2019
2. Verheeij AR, Maas J, Groenewegen PP, De Vries S, Spreuwenberg P (2006) Green space, urbanity and health: how strong is the relation? J Epidemiol Community Health 60(7):587–592. https://doi.org/10.1136/jech.2005.043125
3. Davis JL, Green JD, Reed A (2009) Interdependence with the environment: commitment, interconnectedness, and environmental behavior. J Environ Psychol 29(2):173–180. https://doi.org/10.1016/j.jenvp.2008.11.001
4. Cooper Marcus C, Sachs N (2014) Therapeutic landscapes. An evidence-based approach to designing healing gardens and restorative outdoor spaces. Wiley, New York
5. Thaneshwari KP, Sharma R, Sahare HA (2018) Therapeutic gardens in healthcare: a review. Ann Biol 34(2):162–166. ISSN: 0970-0153
6. Relf PD (2019) Gardens in health care: healing gardens, therapeutic gardens, and horticultural therapy gardens. Acta Hortic 1246:35–40. https://doi.org/10.17660/ActaHortic.2019.1246.6
7. Ulrich RS, Perkins RS (2017) The impact of a hospital garden on pregnant women and their partners. J Perinat Neonatal Nurs 31(2):186–187. https://doi.org/10.1097/JPN.0000000000000247
8. Pedrinolla A, Tamburin S, Brasioli A et al (2019) An indoor therapeutic garden for behavioral symptoms in Alzheimer's disease: a randomized controlled trial. J Alzheimers Dis 71(3):813–823. https://doi.org/10.3233/jad-190394
9. Zimmerman S, Williams SC, Reed SP, Boustani M, Preisser SJ, Heck E, Sloane SP (2005) Attitudes, stress, and satisfaction of staff who care for residents with dementia. Gerontologist 45:96–105. https://doi.org/10.1093/geront/45.suppl_1.96
10. Ulrich RS, Zimring C, Zhu X et al (2008) A review of the research literature on evidence-based healthcare design. HERD 1(3):61–125. https://doi.org/10.1177/193758670800100306
11. Bernez L, Batt M, Yzoard M et al (2018) Therapeutic gardens also offer a valuable setting for burnout prevention. Psychol Fr 63(1):73–93. https://doi.org/10.1016/j.psfr.2017.02.001
12. Ulrich RS (1979) Visual landscapes and psychological wellbeing. Landsc Res 4:17–23. https://doi.org/10.1080/01426397908705892
13. Kaplan R, Kaplan S (1989) The experience of nature: a psychological perspective. Cambridge University Press, Cambridge
14. Kaplan S, Kaplan R (1995) The restorative benefits of nature: toward an integrative framework. J Environ Psychol 15(3):169–182
15. Grahn P, Stigsdotter UA (2003) Landscape planning and stress. Urban For Urban Green 2:1–18
16. Kaplan R (1973) Some psychological benefits of gardening. Environ Behav 5(2):145–161
17. Ulrich R (1984) View through a window may influence recovery from surgery. Science 224(4647):420–421
18. Uwajeh P, Iyendo TO, Polay M (2019) Therapeutic gardens as a design approach for optimising the healing environment. Explore J Sci Heal 15(5). https://doi.org/10.1016/j.explore.2019.05.002
19. Cooper Marcus C, Barnes M (1999) Healing gardens. Therapeutic benefits and design recommendations. Wiley, New York

20. Castaldi S, Bevilacqua L, Arcari G et al (2010) How appropriate is the use of rehabilitation facilities? Assessment by an evaluation tool based on the AEP protocol. J Prev Med Hyg 51(3):116–120
21. Rodiek S, Nejati A, Bardenhagen E et al (2016) The seniors' outdoor survey: an observational tool for assessing outdoor environments at long-term care settings. Gereontologist 56(2):222–233. https://doi.org/10.1093/geront/gnu050
22. Bottero MC, Buffoli M, Capolongo S, Cavagliato E, di Noia M, Gola M et al (2015) A multidisciplinary sustainability evaluation system for operative and in-design hospitals. In: Capolongo S, Bottero MC, Buffoli M, Lettieri E (eds) Improving sustainability during hospital design and operation: a multidisciplinary evaluation tool. Springer, Cham, pp 31–114. https://doi.org/10.1007/978-3-319-14036-0_4
23. Buffoli M, Rebecchi A, Gola M et al (2018) Green soap. A calculation model for improving outdoor air quality in urban contexts and evaluating the benefits to the population's health status. In: Mondini G, Fattinnanzi E, Oppio A, Bottero M, Stanghellini S (eds) Integrated evaluation for the management of contemporary cities. Green energy and technology. Springer, pp 453–467. https://doi.org/10.1007/978-3-319-78271-3_36
24. Brambilla A, Capolongo S (2019) Healthy and sustainable hospital evaluation—a review of POE tools for hospital assessment in an evidence-based design framework. Buildings 9(4):76. https://doi.org/10.3390/buildings9040076
25. Brambilla A, Buffoli M, Capolongo S (2019) Measuring hospital qualities. A preliminary investigation on health impact assessment possibilities for evaluating complex buildings. Acta Bio-Med: Atenei Parm 90(9S):54–63. https://doi.org/10.23750/abm.v90i9-S.8713
26. Faroldi E, Fabi V, Vettori MP, Gola M, Brambilla A, Capolongo S (2019) Health tourism and thermal heritage. Assessing Italian Spas with innovative multidisciplinary tools. Tour Anal 24(3):405–419. https://doi.org/10.3727/108354219X15511865533121
27. Mosca EI, Herssens J, Rebecchi A, Strickfaden M, Capolongo S (2019) Evaluating a proposed design for all (DfA) manual for architecture. Adv Intell Syst Comput 776:54–64
28. Arslan M, Kalaylioglu Z, Ekren E (2018) Use of medicinal and aromatic plants in therapeutic gardens. Indian J Pharm Educ Res 52(4):151–154
29. Capolongo S, Gola M, Di Noia M, Nickolova M, Nachiero D, Rebecchi A, Settimo G, Vittori G, Buffoli M (2016) Social sustainability in healthcare facilities: A rating tool for analysing and improving social aspects in environments of care. Annali dell'Istituto Superiore di Sanita, 52(1):15–23. https://doi.org/10.4415/ANN-16-01-06
30. Gianfredi V, Buffoli M, Rebecchi A, Croci R, Oradini-Alacreu A, Stirparo G, Marino A, Odone A, Capolongo S, Signorelli C (2021) Association between Urban Greenspace and Health: A systematic review of literature. Int J Environ Res Public Health 18:5137. https://doi.org/10.3390/ijerph18105137
31. Mosca EI, Herssens J, Rebecchi A, Capolongo S (2019) Inspiring architects in the application of design for all: Knowledge transfer methods and tools. J Access Des All 9(1):1–24. https://doi.org/10.17411/jacces.v9i1.147

Approaches to Post-Occupancy Evaluation and Wellbeing in Designed Space

Paolo Costa and Leonardo Chiesi

From a sociological perspective, the ex-post evaluation of a designed space is an applied research tool and a learning process that is useful at various degrees: to understand how to improve the specific space that is evaluated; to gather knowledge as an input for the future design of other spaces of the same typology; and, on a higher level, to improve knowledge and sensitivity about the complex relationship between people and space, that should always play a crucial role in every act of design.

1 Two Gaps: Before and After Occupancy

The evaluation of a designed space requires the examination of two perspectives: on one side, it is necessary to investigate how spaces are designed and built; on the other, how spaces are appropriated by their inhabitants.

The process of design starts with a strategic phase, where the general goals of the project and the principles that have to inform it are laid out. Then, in a programming phase, these goals are translated into the specific elements that the project has to include. In the design phase the general and specific goals decided in the previous

P. Costa (✉) · L. Chiesi
Dipartimento di Scienze Politiche e Sociali, Università degli Studi di Firenze, Florence, Italy
e-mail: paolo.costa@unifi.it

L. Chiesi
e-mail: chiesi@unifi.it

Scuola di Architettura, Università degli Studi di Firenze, Florence, Italy

L. Chiesi
Fondazione per il Futuro delle Città, Florence, Italy

phases are transformed into design solutions that should satisfy those goals. Finally, in the construction phase, the solutions are turned into a tangible reality.[1]

In each of these phases, assessing different options is crucial. Each choice made in one phase has advantages and disadvantages that have to be carefully considered, as the quality of the choices made in the first stages has consequences on the quality of those that will be made in the following ones.

Architects and designers know that this description is an abstraction, a rationalization of a process that in reality is less linear. Architectural practice is a field where readiness to react and reconsider decisions that before seemed unquestionable can make the difference, especially in a competitive market. The complexity of this process, the number of the actors involved and their (often contrasting) interests are some of the many factors that make this predictable pattern unlikely to manifest. The typical long duration of the entire process that leads to construction and delivery also increases the chances that the conditions—for example, in terms of resources, regulations, and final recipients—in which the project was initially imagined will be different at the time of its completion.[2] For these reasons, there is always a discrepancy between the initial intentions of the project, declared in the strategic phase, and the intentions that the project can fulfil—although still potentially—at its completion. This discrepancy can be defined as *pre-occupancy gap*.

At that stage, nothing can be still said about how the project will be actually inhabited by final users. In other words, it is still unknown how the set of opportunities and intentions inscribed in it by the designer will meet the intentions of the final users when they will use it. Building on a conceptual pair by Gans [2, p. 6], these two sets of intentions can be defined as *potential space* and *effective space*.

Comparing potential and effective space means to analyse the *post-occupancy gap*. This discrepancy is always present, independently from the effort put by designers: on the one hand because it is impossible to predict *all* the characteristics, the needs, and the practices of the final users; on the other hand because, while these are fluid and change over time, the opportunities inscribed in space by the designer are set and oppose change with their materiality.

As shown in Fig. 1, pre-occupancy gap is internal to the domain of the production of architecture.[3] Post-occupancy gap, instead, is related with the relationship between two domains, that of the built environment and that of the life that, after occupancy, appropriates this environment.

---

[1] This is just one of the many possible models of the design process, adapted from [1].

[2] Despite that, this lack of linearity is rarely told, both in documents that accompany the project at its delivery, nor in publications and magazines that describe the project to the general public. How things really went is often a sacred secret that has to be preserved, for the sake of perfection, as if a plain design process could guarantee a better life of the project after its delivery.

[3] Viewing the conclusion of the production of architecture right after construction is certainly a simplification. Landscape architecture, for example, is more used to plan how plants and other natural elements will grow and develop also after occupancy. Nonetheless, in many design cultures it is still a mainstream approach to consider delivery as a punctual moment, an event, instead of a process that goes beyond occupancy.

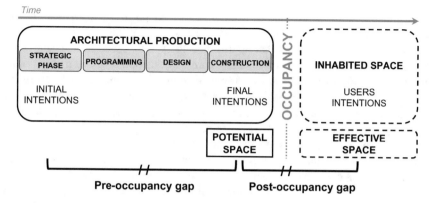

Fig. 1 Pre-occupancy and post-occupancy gaps and their relationship with the phases of architectural production and with inhabited space. *Source* ©Paolo Costa

From a sociological perspective, the evaluation of architecture and design moves from the awareness of the inevitability of the post-occupancy gap. Hence, the main goal of evaluations should be to create knowledge that contributes to minimize that gap.

If the main focus of Post-Occupancy Evaluation (hereafter POE) is certainly on what happens to the project *after* it has been inhabited, its analytical force is also strengthened by the reconstruction of what happened before the occupancy, during the design and construction process. If we want to learn from an issue related to a post-occupancy gap and not just describe it, we have to try to understand its causes. Is it due to a lack of knowledge about the final users? Or to an underestimation of some details that were not as irrelevant as it was thought? Or maybe the issue is caused by a change to the project made in the last stages of the design process, in the construction phase, for example, due to a budget reduction or to a change of regulations. That would make the issue not necessarily related to a bad design choice but to a pre-occupancy gap. What has to be stressed is that searching for causes helps to create knowledge that can be an input in future design processes.

2 How We Live and Inhabit Space

Evaluation as scientific practice has developed in many different directions, with a variety of approaches. As we will see, the epistemological debate on evaluation of built environment and architecture is not as rich as the debate about evaluation of social interventions. One way to help this development is to try to build on a theory on how people inhabit and engage with space. One first key element is that the mutuality of this relationship produces both *place-making* and *emplacement* processes: while we 'make' space and territorialize it, we also do that in ways that are profoundly influenced by the space itself.

To make a step further, one of the authors [3, 4] proposed a tripartite model of how space frames social action, trying to systematize the scientific outcomes of various research programs that approached the people-space relationship from different perspectives. It is worth recalling here the main points of that model, for its strong methodological implications on some theoretical and operative aspects of the practice of POE. According to this model, space offers three classes of opportunities.

The opportunities of the first kind are *affordances*, defined by the psychologist of perception Gibson [5, pp. 127–143] as perceived opportunities for action mediated by the environment. Gibson's idea—later confirmed by evidence emerged from neuroscience[4]—was that we perceive opportunities in space directly, in a way that does not imply a cognitive mediation. This is how we detect *what* and *how* we can do things in the proximate space that surrounds us. In short, what space *affords* us. In this perspective, affordances are not *in* the spatial features: they are perceived by us in our relationship between space and our body and mind. Different people, hence, can perceive different affordances in the same space, according to their bodies or their needs, but also according to what that specific space affords. A bench can be used to seat in the most common normative way, but also to sit astride, or on top of the backrest; to hide behind, to play below, or to display objects.

People can hardly explain why they use space in a certain way, at least for the aspects that are more strongly related to affordances. This is because we don't perceive affordances in space using cognition, but through a continuous 'scanning' process of space around us that our brain performs without a necessary conscious involvement.

The second kind of opportunity relates to signs as references to meaning. Space, in other words, is used to embed signs which express content and subsequently to associate content to signs. This semantic dimension is therefore two-fold. On one side, people encode meanings in space through signs: for example, they express their belonging to a specific culture or social group, or they signal their power or status, through the objects with which they populate their space. At the same time, people are involved in continuous decoding acts of signs that others have encoded in space, no matter if those others are their peers or designers. When considering signs, therefore, space can be construed as a platform for communication of meanings between encoders and decoders [7, 8]. This operation is only possible if the actors of this communication process share the same code that links meanings and signs. This code is a prerequisite for the communication to happen and is a socially constructed and culturally driven mediator. When the link between a sign and a meaning is not familiar to—or it is not the same for—all the actors, issues arise. This is common when we are immersed in a different cultural or subcultural symbolic spatial setting—as travellers know very well.

The third way we relate to space is through the aesthetic experience that we live in it, through our senses. Processes that are activated in this modality of experiencing space allow us to somehow 'taste' it, elaborating a continuous flow of sensorial responses to the stimuli that we receive.

---

[4] For a reconstruction of the use of the concept of 'affordance' in neuroscience, see [6].

In some cases, we taste space *actively*, so we feel our sensorial involvement, to the point that we feel if we want more or, on the contrary, less of that specific sensorial experience [3, pp. 20–24]. This is what happens to us, for example, when we enter a room that we feel too dark, or we sit on a very cold surface, or we walk bare feet on a lawn or a sandy beach; or when we search for shadow. But there are also cases in which we 'taste' space *passively*. That means that, despite the fact that we don't feel any particular stimulus happening, the environment has an impact on us nonetheless. This happens, for example, in all cases in which marketing experts and psychologists play with the environment to influence our consuming behaviours, through light, music, smell, etc. [9].

This modality is also the one in which much of our generally positive responses to natural and green environments take place—both actively and passively. A response that is strongly linked to the deep relationship that our species had for such a long time with the natural environment [10].

It needs to be emphasized that people experience opportunities mediated by space without acknowledging if they are related to affordances, meaning or taste. In many cases, in fact, our behaviour in space is the result of these modalities fluidly blending as one. For example, the reason that could drive a patient to sit in a specific part of a hospital garden could be related to the fact that, at the same time, the seating allows her to talk face to face to the friend who visited her (affordance); but also because it reminds her of her grandmother's backyard (meaning); and also because she is unconsciously attracted by the smell of the plants that surround that spot (taste).

As it is clear, capturing the complexity of these partially unconscious processes is not an easy research challenge. However, the analytical deconstruction of such processes with this tripartite model is helpful to thoroughly investigate what people do *in* space and *with* space. But, as we will see, it also helps to choose between different approaches to evaluation of the relationship between people and space.

3 Three Approaches to Post-occupancy Evaluation

One of the authors [11] proposed a typology of the three main approaches that have shaped the field of Post-Occupancy Evaluation and of the studies on the relationship between space and people. This typology was built analysing these practices since the second half of the XX century, when research on this relationship started to spread in various forms in the scientific community, primarily in United States.[5] Recalling the main features of the three approaches will help to discuss some current challenges of the practice of POE.

---

[5] This typology is based on the analysis of the main approaches to evaluation of social interventions developed by Stame [12]. From a sociological perspective, in fact, a design project can be intended as a social program that uses spatial devices to achieve its goals.

3.1 The Experimental Approach

The first approach is mainly rooted in the field of environmental psychology, that in the 1950s started to analyse the effects of different spatial features (such as layout, colours, materials, etc.) on people inhabiting spaces—often patients in healthcare spaces.[6] Most of these studies were inspired by a neo-positivist research approach, typical of mainstream psychology, based on the idea that psychology—and other social sciences—should adopt the same research methods and tools of natural sciences: the experimental research design. In short, a typical design of these experiments—or quasi-experiments—would define an initial hypothesis about the effects of a spatial change on a specific psychological or social dimension, for example of a new bedroom layout on the social attitude of patients of a mental ward. Then, it would test that hypothesis by measuring that psychological dimension before and after the spatial intervention—considered as a stimulus—with questionnaires and tests that produced quantitative results. Although these studies were not explicitly labelled 'ex-post evaluations', they established a modus operandi to approach the relationship between people and space that has a strong impact on evaluation research today.

3.2 The Social Construction of Design Success

Another approach to evaluation developed in the 1960s, building on the works of sociologists and geographers who focused on the relationship between people and space from a novel perspective (see e.g. [15, 16]). In these evaluations a much stronger voice was given to people who inhabited spaces—often housing projects for low income families in the USA—than it traditionally happened in current assessment of architectural quality.[7] Furthermore, these studies built on the idea that it is impossible to define a-priori goals for architecture; and that there isn't any "one best way" to design the environment that can be applied to all contexts.[8] This meant that the success of a design project was not necessarily the achievement of the set of goals defined at the beginning of the design process. Rather, the definition of what would be a successful design had to be socially constructed, involving not only designers but also inhabitants and all the other actors that acted in the development of the project or engaged with it after its completion.

[6] See, for example, the classic study of Osmond [13], in Canada, or Proshansky et al. [14], in USA.

[7] As an example, many architectural prizes were given without consulting people that inhabited the projects, or before they were inhabited or even built. That practice—still very common today—clearly shows a very narrow way of conceiving architectural quality.

[8] The idea of the "one best way", in fact, was typical of the Modern movement in architecture and of Functionalism in planning, that had been mainstream for decades, but that in those years were starting their decline.

This constructivist approach looked at the project in a processual perspective. While the experimental approach considered the spatial intervention as a stimulus, a finite event, and focused only on what came before and after it, social scientists deconstructed the design process investigating the assumptions that connected each design choice to its expected goals. Then, they studied if the delivered project was realized according to such goals. Lastly, they analysed how inhabitants related to those design choices. The methodological tools used in these evaluations included structured tools, such as questionnaires; but they also much relied on a wide set of unstructured research tools, such as in-depth interviews, focus groups, and sessions of observation.

These studies uncovered a variety of ways in which people relate to spaces, showing the social dimensions that affect such relationships. They also acknowledged the relevance of the unexpected outcomes of design—not necessarily negative for the project. And in many cases they proposed sets of guidelines for future projects.

3.3 Performance and Quality

The third approach to evaluation of spaces consolidated in the mid 1980s, building on the pragmatist framework to evaluation of social interventions and policies [17]. Given the background in engineering of its main actors, the evaluations of this approach were initially concerned mainly with the technical performance of the built environment.[9] After this initial phase, the scope of such evaluations widened, at least in their declared goals: first, the evaluation should not start only after the construction, but should also be implemented during the architectural production, considering each of its stages; second, the evaluation should be concerned with all kinds of performance of the built environment, including the social and psychological aspects.

Since then, this approach to POE has been widely practiced. The relatively simple design strategy of the evaluation also helped: for this approach, evaluating would mean to measure a set of dimensions in several cases; then, to build a series of standards related to those dimensions; finally, experts could use those standards to benchmark further case studies. The creation of standardized evaluation tools guaranteed—at least in principle—replicability, which made this approach even more appealing.

3.4 Evolution and Success of the Approaches

These approaches have contributed in different ways to the advancement of Post-Occupancy Evaluation practice and to the development of the body of knowledge

---

[9] See, for example, [1, 18].

about the relationship between people and space. These approaches have evolved along different paths, with different degrees of popularity among practitioners, and with different implications in regards with wellbeing in designed spaces.

The *experimental approach* currently constitutes the research framework of Evidence-Based Design (EBD),[10] that posits that design choices should be based on scientific evidence. EBD is strongly influenced by the Evidence-Based Medicine paradigm (EBM),[11] that since the 1990s introduced in medicine the idea that diagnostic and therapeutic practices should be validated by robust quantitative studies, based on randomized-control trials. The growing success of EBD in a field—architecture—that is not used to justify its choices is promising. However, the strict adoption of the same quantitative methodological toolkit of EBM has been criticised for its incapacity to grasp the complexity of some latent but relevant dimensions that structure the people-space relationship.[12]

The *constructivist approach*, with the involvement of all stakeholders and its effort to deconstruct the design process to put it in relationship with the life that takes place in spaces after occupation, has developed in the most analytic kind of ex-post evaluation, often defined as 'diagnostic evaluation'.[13] This approach allows the highest expression of the heuristic potential of Post-Occupancy Evaluations: its research tools are particularly suitable for the development of hypotheses about the causes of success and failure of design choices, both by exploring the pre-occupancy and the post-occupancy gaps. However, the success of diagnostic evaluation has been limited: the resources needed to design and execute the evaluations is considered hardly compatible with the fast rhythm of mainstream architectural practice.

Lastly, the *performance approach*, with its typical standardised tools, has been considered to be particularly effective in the definition of standards of quality for many architectural typologies. On one side, those standards facilitate the comparison among case studies; on the other, they are particularly helpful to improve quality in typologies of spaces with public functions, for which evaluation is a functional equivalent of what competition is in the private sector.

4   Evaluation for Wellbeing: Current Challenges

Connecting the topics discussed in this chapter—the acknowledgement of the two gaps before and after occupancy, the theoretical framework about habitation, and the approaches to Post-occupancy Evaluation—is useful to point out some of the

---

[10] See, e.g., [19], and the monographic issue of "Environment and Behavior" edited by Zimring and Bosch [20].

[11] On EBM, see [21, 22].

[12] On the relevance of EBD but also on its issues see, e.g., [23, p. 16]; for a strong criticising position of EBD see [24].

[13] Post-Occupancy Evaluations have been defined as *indicative, informative* or *diagnostic*, each respectively with a higher degree of in-depth analysis, but also of resources needed to perform them [25].

challenges that currently affect the practice of POE and to offer advice to overcome them.

As seen in the analysis on habitation, the three dimensions are all relevant to understand people's relationship with space. However, most evaluations still focus mainly on the most technical and measurable aspects of this relationship. This under-estimates the relevance of other aspects that, despite not being measurable, have a crucial role in defining human experience in space. Many social and cultural aspects that affect this relationship require to be investigated with indirect and qualitative strategies. For example, the domain of habitation that is related to meanings requires research strategies that cannot be limited to quantitative and structured tools.

The same can be said about the analysis of the opportunities in space related to affordances. As we have seen, the concept of 'affordance' is particularly fertile to investigate what people do in space and with space and, specifically, how different people inhabit the same space perceiving different opportunities in it. However, connecting this construct with a proper methodological toolkit to study affordances is difficult. Limiting to interviewing people about what they do in space can hardly grasp the latent reasons for their behaviours.[14] Observation is a much more effective strategy, given that much of our behaviour related to affordances is mostly unreflective and relies on tacit understanding.

Strategies to assess spaces in terms of affordances have been recently proposed, for example, in the field of therapeutic landscape evaluation.[15] These evaluation strategies often adopt tools that are typical of the performance approach (e.g. audit, scorecards, rating scales, etc.) that wisely assess how specific spatial devices afford certain actions to specific groups of people. This drive to specificity is crucial: as said above, different actors do not necessarily perceive the same set of affordances in the same space. Hence, it is very important to limit the assumption—sometimes persistent in some advocates of the performance approach—that it is possible to develop evaluative tools that are universally valid. On the contrary, for example, assessing healing gardens in children's hospitals is very different from assessing such spaces in centres for people with Alzheimer's disease. This diversity of users and spaces needs to reflect in the development of diverse and specific assessing tools.

The act of assessing if certain spatial features allow a specific activity should be based on the hypothesis that those activities are needed and desired by the specific group of users that inhabit that space. These needs and desirability, though, should have always been previously validated with users, a step that is too often neglected. The same goes for the assumptions regarding how specific spatial features or design choices actually promote processes that are consistent with the expected goals of a space: such as, for example, fostering social interaction among peers or, on the contrary, allowing a certain level of privacy.

As Wood [29, p. 121] has pointed out in his analysis of POE of schools, the particular context, people, aims and values involved in evaluative tools and exercises

---

[14] For an ex-post evaluation that shows the relevance of these latent dimensions in the people-space relationship, see [26].

[15] See, for example, [27, 28].

matter more than is generally recognised. Every evaluation should be based on situational knowledge contingent on the specific relationship between people and space that is being assessed. What needs to be stressed is that this knowledge is more effectively produced by the heuristic potential of the in-depth and unstructured tools. This knowledge can then be complemented by structured tools—such as questionnaires—when the object of study is large enough to necessitate quantitative descriptions to better capture its complexity. We can easily count what people do in space and find that a specific design solution produces certain effects more than others. But if we don't understand why that happens—or, in other words, how the three domains of experience in space are involved—we can hardly reproduce a successful design solution (or avoid a bad one) in another space, where the overall context and the users will be inevitably different in many respects.

Overcoming these challenges requires a stronger theoretical framework and methodological reflexivity on Post Occupancy Evaluation than before. This sets a path for the development of a practice that, we believe, still has a lot to contribute to the design of better spaces for the wellbeing of people.

References

1. Preiser WFE, Schramm U (1997) Building performance evaluation. In: Watson D, Crosbie MJ, Callender JH (eds) Time-saver standards for architectural design data, 7th edn. McGraw-Hill, New York
2. Gans H (1968) People and plans. Essays on urban problems and solutions. Basic Books, New York
3. Chiesi L (2014) Situarsi nello spazio progettato. Per una definizione sociologica dell'abitare. In: Chiesi L, Surrenti S (eds) L'ospedale difficile. Lo spazio sociale della cura e della salute. Liguori, Napoli, pp 1–26
4. Chiesi L (2016) Territoriality as appropriation of space. How «engaging with space» frames sociality. In: Dessein J, Battaglini E, Horlings L (eds) Cultural sustainability and regional development. Theories and practices of territorialisation. Routledge, London, pp 73–94
5. Gibson JJ (1979) The ecological approach to visual perception. Boston: Houghton Mifflin
6. de Wit MM, de Vries S, van der Kamp J, Withagen R (2017) Affordances and neuroscience: steps towards a successful marriage. Neurosci Biobehav Rev 80:622–629
7. Lawson B (2001) The language of space. Architectural Press, Oxford
8. Rapoport A (1982) The meaning of the built environment. A nonverbal communication approach. Sage, Beverly Hills, CA
9. Anthony KH (2017) Defined by design: the surprising power of hidden gender, age, and body bias in everyday products and places. Prometheus Books, Amherst, NY
10. Kuo M (2015) How might contact with nature promote human health? Promising mechanisms and a possible central pathway. Front Psychol 6
11. Costa P (2014) Valutare l'architettura. Ricerca sociologica e post-occupancy evaluation. FrancoAngeli, Milano
12. Stame N (2001) Tre approcci principali alla valutazione: distinguere e combinare. In: Palumbo M (ed) Il processo di valutazione. Decidere, programmare, valutare. Franco Angeli, Milano, pp 21–46
13. Osmond H (1957) Function as the basis of psychiatric ward design. Ment Hosp 23–29
14. Proshansky HM, Ittelson WH, Rivlin LG (eds) (1970) Environmental psychology: man and his physical setting. Holt, Rinehart, and Winston, New York

15. Cooper Marcus C (1975) Easter Hill Village. Some social implications of design. The Free Press, New York
16. Zeisel J, Griffin M (1975) Charlesview Housing. A diagnostic evaluation. Harvard University, Cambridge, MA
17. Scriven M (1980) The logic of evaluation. Edgepress, Inverness, CA
18. Preiser WFE, Vischer J (eds) (2005) Assessing building performance. Elsevier, Oxford
19. Ulrich RS, Zimring CM, Zhu X, DuBose JR, Seo HB, Choi YS, Quan X, Joseph A (2008) A review of the research literature on evidence-based healthcare design. Health Environ Res Des J 1(3):101–165
20. Zimring C, Bosch S (eds) (2008) Building the evidence base for evidence-based design. Environ Behav [monographic issue] 40(2)
21. Guyatt G (1992) Evidence-based medicine: a new approach to teaching the practice of medicine. J Am Med Assoc 268:2420–2425
22. Sehon S, Stanley D (2003) A philosophical analysis of the evidence-based medicine debate. BMC Health Serv Res 3(1):14
23. Cooper Marcus C, Sachs N (2014) Terapeutic landscapes. An evidence-based approach to designing healing gardens and restorative outdoor spaces. Wiley, Hoboken, New Jersey
24. Stankos M, Schwarz B (2007) Evidence-based design in healthcare: a theoretical dilemma. Interdiscip Des Res e-J I(I: Design and Health)
25. Preiser W, Rabinowitz HZ, White ET (1988) Post-occupancy evaluation. Van Nostrand Reinhold, New York
26. Costa P, Laurìa A, Chiesi L (2020) Promoting autonomy through home adaptations. Appropriation of domestic spaces in Italy. Disabil Soc 1–24
27. Bardenhagen E, Rodiek S (2016) Affordance-based evaluations that focus on supporting the needs of users. Health Environ Res Des J 9(2):147–155
28. Cooper Marcus C (2007) Alzheimer's garden audit tool. J Hous Elder 21:179–191
29. Wood A (2018) The politics of post occupancy evaluation: the example of schools. In: Alterator S, Deed C (eds) School space and its occupation: conceptualising and evaluating innovative learning environments. Brill/Sense, Leiden, pp 121–134

A "Prosthetic Environment" for Individuals with Dementia

Antonio Guaita

1 Introduction

Dementia and cognitive impairment[1] are among the leading causes of disability and dependence in the elderly. They place a considerable burden on families and caregivers and are becoming a major challenge for healthcare systems generally, as well as for society at large [20, 42, 43]. To date, there are no pharmacological treatments that can stop or reverse the clinical course of the disease [4].

Although dementia is primarily defined as a form of cognitive impairment (Box 1), it is the non-cognitive and behavioural symptoms characterising the protracted moderate-severe phase of the disease that are the greatest source of distress both for sufferers and for their caregivers.

Box 1

DSM V: from "dementia" to "Major Neurocognitive Disorders" (the alternative term was prompted by a desire to address the stigma associated with the term 'dementia').
Diagnosis of Major Neurocognitive Disorders requires:

1. evidence of significant cognitive decline from a previous level of performance in one or more of the six cognitive domains detailed in the Manual:

    - Complex attention: sustained attention, divided attention, selective attention and information processing speed
    - Executive function: planning, decision making, working memory, responding to feedback, inhibition and mental flexibility

---

[1] The term "dementia" refers to a syndrome, i.e. a set of symptoms, while Alzheimer's disease (AD) is a disease, the main cause of dementia. This is why in this chapter AD will often be used to represent dementias in general.

A. Guaita (✉)
Fondazione Golgi Cenci, Abbiategrasso, Italia
e-mail: a.guaita@golgicenci.it

© The Author(s), under exclusive license to Springer Nature Switzerland AG 2023
S. Capolongo et al. (eds.), *Therapeutic Landscape Design*,
PoliMI SpringerBriefs, https://doi.org/10.1007/978-3-031-09439-2_7

- Learning and memory: free recall, cued recall, recognition memory, semantic and autobiographical long-term memory, and implicit learning
- Language: object naming, word finding, fluency, grammar and syntax, and receptive language
- Perceptual-motor function: visual perception, visuoconstructional reasoning and perceptual-motor coordination
- Social cognition: recognition of emotions, theory of mind and insight

2. The cognitive deficits are sufficient to interfere with independence (i.e. requiring minimal assistance with instrumental activities of daily living).
3. The cognitive deficits do not occur exclusively in the context of a delirium.
4. The cognitive deficits are not primarily attributable to another mental disorder (for example major depressive disorder and schizophrenia).

Source: American Psychiatric Association. Diagnostic and Statistical Manual of Mental Disorders, 5th ed. Washington: American Psychiatric Association; 2013.

Approximately five out of every six patients with dementia, including those living at home, will develop behavioural and psychological symptoms of dementia (BPSD) during the course of the disease [1, 18, 23].

Agitation and other BPSD hinder activities and relationships, cause feelings of helplessness and distress in families and formal caregivers, and are strong predictors of poor quality of life [16] as well as nursing home admission [17].

Many of these symptoms cannot be attributed solely to the disease, but also to failure to adapt the living environment to dementia sufferers' altered and reduced ability to understand the meaning of their surroundings [15] (Fig. 1). After all, no one has ever built cities and towns, hospitals, care homes and private homes with the needs of dementia sufferers in mind. Consequently, as the disease progresses, the physical environment surrounding affected individuals increasingly becomes, for them, an unfamiliar and sometimes even dangerous place [28].

The presence of dementia raises a problem, for care settings, that clearly distinguishes the disability of those affected by the disease from the "disability" that we typically associate with frail, dependent elderly people.

Indeed, the disability of dementia sufferers is characterised not by impaired or reduced movement, but rather by their tendency to move about and wander a great deal, too much even, often without really knowing where they are going [6, 8]. It is therefore necessary that their environment furnish them, from the outside, with

Fig. 1 The human behaviour is never the direct product of the pathology, but the result of the interaction between pathology and environment demands

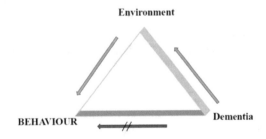

elements able to give them the sense of security and the ability to understand space that can no longer come from within: this is the reasoning underlying the "prosthetic environment" concept.

A lower limb amputee, given that no amount of exercise will ever restore his lost limb, needs to be given a prosthetic limb in order to regain, at least to an extent, the ability to walk. Similarly, since dementia is the result of irretrievable loss of a part of the brain (and mind), we should try to compensate, artificially, for this lost portion of the brain. This can be done thorough the construction of a complex external "prosthesis" made up of at least three components: people, activities and physical space [32].

In other words, to be suitable for dementia sufferers, an environment must be more than just devoid of architectural barriers. Above all, it must have "prosthetic" characteristics of the kind outlined in an article published in JAMA [19]:

An individual with degenerative advanced impairment of the brain, such as severe Alzheimer's disease, whose brain is reduced in size and number of neurons and synapses [...] needs external support for maintaining at least some cognitive function. The key theoretical concept is that individuals with dementia can obtain from external sources what they cannot from inside. The usual definition of prosthesis ("artificial part supplied to replace a missing body part" Trumble WR, ed. *Oxford English Dictionary*. New York, NY: Oxford University Press; 2007) is limiting, because physicians use several prosthetic interventions in their practice that do not necessarily replace a structural part of the body, such as the use of insulin for individuals with diabetes. The prosthetic model for dementia identifies deficits in function in the patient and builds a "prosthesis of care" for each individual that is intended to compensate for the lost function(s). The main goal of the prosthesis is not to regain cognition or function, but to deal with the wellbeing of the person, to achieve the best status in absence of distress and pain. To potentially help brain function, a complex prosthesis is needed, made up of 3 basic elements: the individuals with whom the person with dementia interacts, the physical space in which the person lives, and the programs and activities in which the person engages. Most disturbed behaviors and functional impairments can be considered as the expression of an imbalance between the abilities of the individual with dementia and environmental demands, and not merely the expression of the pathology. Therefore, even if it is not possible to enhance the cognitive function of individuals with dementia, it may be possible to reduce their distressing symptoms through a change of the environment, and an informed approach by others with whom the individual with dementia interacts. Physical space is recognized to influence the behavior of individuals with dementia and a specially designed environment has been proposed [14]. In the prosthetic environment, the most important features are safety, comfort, and access, rather than stimulation. For instance, if an individual with dementia paces or wanders, this person needs a safe environment rather than devices or interventions that prevent walking.

This chapter looks at the principles that should guide the creation of a prosthetic environment designed to increase the wellbeing of people with dementia, and also outlines various sensory abnormalities that, in addition to cognitive changes, underlie the altered perception of physical space in dementia sufferers.

2   General Features of Prosthetic Environments

Typically, an environment tailored to individuals with dementia should be safe, familiar, flexible and comfortable. Furthermore, the purpose of the environment should be rendered unambiguous in a way that takes into account the specific visual abnormalities that frequently accompany dementia. In short, it should be a simple, homely environment and not a high-tech one [5, 7]. Sometimes, in order to meet conflicting needs, such as safety and freedom, high-tech solutions may be required. In this case it is important to make sure they are unobtrusive (e.g. choosing height-adjustable beds with wooden frames, silent alarm systems, and so on) [3].

Safety

Every environmental safety measure adopted increases the dementia sufferer's range of freedom and reduces the need to supervise and confine him/her (which obviously means less stress both for the individual concerned and for the caregiver). Indeed, because of the cognitive changes associated with the disease, in particular the impairment of critical thinking and judgement, individuals with dementia, especially in the advanced stages of the disease, tend to put themselves in situations of risk. Therefore, caregivers frequently start to feel the need to keep them closely monitored. Safety is an issue both for dementia sufferers in residential accommodation and in those who live at home. Residential and (when possible) home environments should be designed, both internally and externally, to be safe, protected areas that allow the individual to enjoy maximum freedom [34]. A useful parallel—even though the two situations are not truly superimposable—can be drawn with the concepts behind the design of areas specifically intended for children, which, it is widely recognised, should promote motor skills and exploration of the environment, while at the same time protecting their safety. In this case, too, the safety measures adopted should be discreet: noisy alarms and wearable sensors are avoided in favour of technology that is less obtrusive and, if necessary, disguised. All choices relating to dementia sufferers' safety should be made taking into account the various ethical issues that arise when the need to respect an individual's right to self-determination is complicated by the fact that the person in question has a disease that, to varying degrees, undermines the capacity for self-determination. Furthermore, it is always necessary to consider the ethicality of the possible alternatives to the single measures that have been, or may be, adopted. Finally, it should be remembered that although safety constitutes a right, its pursuit should never be at the expense of a meaningful life; on the contrary, safety measures should always have the effect of boosting independence and enhancing personal dignity.

Flexibility, versatility

A person with dementia is changing all the time, with the result that a seemingly ideal solution can sometimes become entirely unsuitable within a matter of weeks. Dementia is a long-term condition; the needs of the affected person change during its course, and the environment must be flexible enough to embrace and meet these new needs [11]. Again, it may be useful to draw a parallel with what happens in childhood: one need only think of the way a child's bedroom changes as he grows, and of how, over time, we naturally modify this environment (the furnishings and the available space) according to his changing functional abilities.

Comfort

The purpose of a prosthesis is not to treat, but to improve wellbeing. Accordingly, a prosthetic environment must be not only safe, but also be a pleasant place in which to be [27]. In residential settings, inhospitable and uninviting environments will not be used spontaneously, making continuous staff intervention necessary in order to help people to feel comfortable and "at home". Bearing this in mind, and also the way in which places are closely associated with people and activities, it is advisable to provide carefully chosen environmental "infrastructures", in other words, to furnish the environment with everyday objects or interesting items that will encourage activities and personal interaction (most private homes already have plenty of these environmental infrastructures: books, magazines/newspapers, work baskets, ornaments, telephones, wardrobes, drawers, kitchen utensils, fruit bowls, coat hangers, paintings, posters and so on).

Clarity of purpose

To avoid socially inappropriate behaviours, it is crucial to ensure that the environment does not give out ambiguous messages (for example, an aseptic-looking medical examination room, perhaps tiled and white, may be mistaken for a bathroom/toilet, with predictable consequences).

3   Sensory Abnormalities in Individuals with Dementia: Vision

Vision is one of the main channels through which we gather information about our world. Visual impairment undermines the independence of the elderly [39, 40] and forces them to implement elaborate strategies at the level of sensory processing and cerebral organisation. Dementia sufferers, however, lack the efficient cognitive resources necessary in order to do this (Fig. 2).

This is certainly not an inconsequential problem. Complex visual disturbances, in addition to the fairly rare "visuospatial variants" [29], are seen in many individuals with Alzheimer's disease (AD)—60% have visual disturbances different from those present in aging without dementia [31]—and they can even be present in the very early stages of cognitive decline [25]. Indeed, contrary to what was previously believed, visual disturbances appear early in the course of dementia. Loss of visual function correlates with cognitive decline, which, in AD, has been found to be associated with specific visual disturbances that can contribute to functional difficulties and loss of the capacity to interact with the environment [30].

Fig. 2 The progressive impairment of space representation is well highlighted by the copy of a drawn made in subsequent times by a person with dementia. Author's modification from: [32]

Listed below are the main visual problems generally agreed to occur in this setting [10]:

- reduced contrast sensitivity
- impaired movement perception
- difficulty in identifying colours
- impaired use of light.

3.1 Reduced Contrast Sensitivity

Visual contrast sensitivity is measured as the lowest intensity value at which an object can be distinguished from its surroundings. Impaired contrast sensitivity has been found in 20–32% of AD sufferers and less than 1% of control subjects [33]. Performance, on face recognition tasks, for example, but also in activities of daily living (ADL), is enhanced by increasing the size of images and the level of visual contrast [9]. Reduced contrast sensitivity correlates with pathology of the visual extrastriate, inferotemporal and posterior parietal cortical areas of the brain.

This visual deficit interferes with many ADL, including dressing, bathing and eating. It cannot be modified by glasses but can benefit from environmental changes. A study in cognitively unimpaired elderly individuals demonstrated the importance of this function, showing that 57% of the variation in dependency in performing ADL could be attributed to problems of visual acuity and contrast sensitivity, while a two-fold reduction in contrast sensitivity led to a between three- and five-fold increase in difficulty performing ADL [39, 40].

3.2 Impaired Movement Perception

Literature data on impaired movement perception in AD are conflicting, probably to an extent because of differences, between studies, in the size, speed and direction of the visual stimuli employed; it should also be remembered that common eye conditions, such as glaucoma, can cause similar deficits. Nevertheless, more recent research seems to confirm the possible occurrence of impaired optic flow analysis even in quite early stages of the disease, linking it to a disorder of extrastriate visual cortical motor processing [22].

It is also necessary to consider the possible role of inability to perceive depth and distance, which can develop early in the disease and be independent of the presence of other visual disorders. This problem, together with impaired movement perception, has a major impact in terms of driving errors and road accidents, and also affects the ability to "navigate" domestic space [26]. In a simulated driving task, 33% of individuals affected by AD failed to avoid collisions and accidents, compared with 0% of healthy age-matched controls [35].

3.3 Difficulty in Identifying Colours

Colour discrimination is the ability to identify different colours, which lie on one of two "axes": red-green or blue-yellow. Although there have been studies in which the capacity to discriminate colours was not found to be impaired either in healthy elderly or in AD subjects, the majority of studies report colour discrimination deficits in AD, especially for colours along the blue-green spectrum [36], thereby confirming that the deficit mainly involves colours with lower wavelengths, such as blue. However, no correlation has been found between severity of dementia and colour discrimination ability, given the considerably variable patterns shown by affected subjects: indeed, in some, the main brain area involved in seeing colour, which is located in the extrastriate cortex, near to the area involved in "contrast sensitivity", is affected early in the disease, and in others later on. Impaired colour discrimination is also found, albeit to a milder degree, in normal aging and in ocular diseases (both common ones, like glaucoma, and rarer ones, such as retinitis pigmentosa). Impaired perception of white, which is the sum of all colours, can result in difficulty discerning shapes that require an analytical use of vision (in this sense, the presence of doctors' white coats may impair the simultaneous perception of colour, shape and movement).

The hue, brightness and saturation of colour (the latter being its intensity, which will correspond to that of white, grey and black: e.g. pink, red and crimson) all influence colour vision. Light reflected from surfaces influences, depending on its type and quantity, the extent to which colour is perceived. An object's perceived colour may therefore vary according to the surfaces that surround it (Fig. 3).

The use of highly contrasting colours can enhance the autonomy of people with dementia, and has been found to impact on the daily lives of individuals affected by

Fig. 3 Look at the above figure and hide the below one: the left central square appears darker. Now look at the figure below: the color of the two squares is identical. Author's elaboration

AD. For example, using high-contrast red tableware at midday and evening meals was found to increase the quantity of food consumed by individuals with advanced AD (by 25% and 24% respectively); intake of liquids was also increased (by 61 and 105%). A year later, the use of high-contrast blue tableware was found to replicate the increases obtained with the high-contrast red type, whereas no increase in consumption was associated with use of light-coloured tableware. These findings show that the positive results recorded in the study group were mediated by the fact of using objects with a high visual contrast, and not by their specific colour [13].

3.4 Lighting

Finally, with regard to the "impaired use of light", it needs to be appreciated that on reaching adulthood, we start to require a greater intensity of light in order to continue to perceive visual stimuli correctly. Indeed, to retain this ability, we would need the intensity of the light we receive to double every 13 years after the age of 20. Moreover, at the age of 60 years, we receive 40% less light than we did when we were 20 years old [21]. Alongside this, it is also necessary to consider that elderly people and, in particular, those suffering from AD, frequently present disturbed sleep–wake patterns. Efforts to restart the circadian system by means of exposure to light (or melatonin) seem promising [37]. However, increasing the intensity of daytime light was not found to be effective in reducing other behavioural disorders [2].

4 Examples of Adaptation of Specific Environments

Gardens: Outdoor areas, such as gardens, can provide effective prosthetic support. A setting comprising natural and living things is a valuable source of appropriate stimuli and, for some people, can have relaxing (stress-relieving) connotations. Gardens should be planned with the same attention to safety as indoor areas (e.g. choosing non-stinging, non-poisonous plants). It is important to keep the layout simple and ensure that it provides the full variety of information that a garden can offer (colours, scents, tactile sensations, evidence of seasonal changes, etc.) [38, 41].

Bathrooms. Bathrooms are tricky places, characterised by a high density of obstacles and technological objects, often in too small a space. Furthermore, in nursing homes, which tend to be chronically short of space, they often double up as storerooms, and thus become even more inhospitable. Furthermore, with toilet facilities often arranged in rows of cubicles, or having doors that open directly onto corridors, it becomes difficult to ensure that individuals who have problems being behind closed doors enjoy adequate privacy. Bathrooms need to become more restful and user-friendly places. Temperature is another important issue, as they are often too cold (many elderly dementia sufferers have difficulty regulating their body temperature and often feel the cold more than other residents, although some, on the contrary, may tend to feel uncomfortably hot and therefore keep undressing themselves). In many cases, en-suite bathrooms would be the ideal solution, not only because they can be personalised, but also because it would become easier to help the dementia sufferer to associate the bathroom with getting dressed in the morning and undressed at night, and thus to establish a link with sleep patterns, too.

Day rooms/living rooms. In residential settings the day room (or living room) is often non-descript (a cross between a waiting room and a canteen), whereas the corresponding area in a private home (i.e. the living room) is at the heart of domestic life. Day rooms should be decorated and furnished to be immediately recognisable, and broken down into smaller, more intimate, spaces. One way of doing this is to use furniture as a discreet form of partitioning. Just as importantly, access to this area should be facilitated; preferably, it should be designed as a natural continuation, and widening, of a logical path (rather like a town square in relation to its surrounding streets). It could also be useful to locate it near the kitchen, so as to make it even more easily recognisable as a dining room [24].

Corridors. Corridors, which are naturally ambiguous and impractical areas, should be filled with infrastructures and, if appropriate, turned into areas in which to stroll. Indeed, the Report of the Quality Standards Subcommittee of the American Academy of Neurology recommends, in addition to the "provision of exterior space" and "changes in the bathing environment", that corridors be remodelled "to simulate natural or home settings" [12].

5 Conclusions

Although traditional drug treatments are generally ineffective, showing poor results, this does not mean that nothing can be done for dementia sufferers. Nowadays, these individuals' quality of life can be enhanced far more by modification of their living space than by clinical measures. This chapter has pointed out the many ways in which we can increase their autonomy and alleviate their stress; it has also shown that if our aim is to create an environmental prosthesis that will boost their wellbeing, and that of their caregivers, then we have plenty of scope for doing so.

References

1. Appelhof B et al (2017) The determinants of quality of life of nursing home residents with Young-Onset dementia and the differences between dementia subtypes. Dement Geriatr Cogn Disord 43(5–6):320–329. https://doi.org/10.1159/000477087
2. Barrick AL et al (2010) Impact of ambient bright light on agitation in dementia. Int J Geriatr Psychiatry. https://doi.org/10.1002/gps.2453
3. Brawley EC (2001) Environmental design for Alzheimer's disease: a quality of life issue. Aging Mental Health. https://doi.org/10.1080/713650005
4. Cappa SF (2018) The quest for an Alzheimer therapy. Front Neurol. https://doi.org/10.3389/fneur.2018.00108
5. Chaudhury H et al (2018) The influence of the physical environment on residents with dementia in long-term care settings: a review of the empirical literature. Gerontologist. https://doi.org/10.1093/geront/gnw259
6. Cipriani G et al (2014) Wandering and dementia. Psychogeriatrics 135–142. https://doi.org/10.1111/psyg.12044
7. Colombo M, Vitali S, Molla G, Gioia P, Milani M (1998) The home environment modification program in the care of demented elderly : some examples. Arch Gerontol Geriatr 6(supplement):83–84
8. Colombo, M. et al. (2001) 'Wanderers: Features, findings, issues', Archives of Gerontology and Geriatrics, 33(SUPPL.). doi: https://doi.org/10.1016/S0167-4943(01)00127-3.
9. Cronin-Golomb A et al (2000) Facial frequency manipulation normalizes face discrimination in AD. Neurology. https://doi.org/10.1212/WNL.54.12.2316
10. Cronin-Golomb A, Hof PR (2004) Vision in Alzheimer's disease. Gerontologist 35(3):370–376. https://doi.org/10.1093/geront/35.3.370
11. Delva F et al (2014) Natural history of functional decline in Alzheimer's disease: a systematic review. J Alzheimer's Disease: JAD. https://doi.org/10.3233/JAD-131862
12. Doody RS et al (2001) Practice parameter: management of dementia (an evidence-based review): report of the quality standards subcommittee of the American Academy of Neurology. Neurology. https://doi.org/10.1212/WNL.56.9.1154
13. Dunne TE et al (2004) Visual contrast enhances food and liquid intake in advanced Alzheimer's disease. Clin Nutr. https://doi.org/10.1016/j.clnu.2003.09.015
14. Friedrich MJ (2009) Therapeutic environmental design aims to help patients with Alzheimer disease. JAMA—J Am Med Assoc. https://doi.org/10.1001/jama.2009.809
15. Garre-Olmo J et al (2012) Environmental determinants of quality of life in nursing home residents with severe dementia. J Am Geriatr Soc 60(7):1230–1236. https://doi.org/10.1111/j.1532-5415.2012.04040.x
16. Gatz M et al (2003) Sex differences in genetic risk for dementia. Behav Genet 33(2):95–105. https://doi.org/10.1023/A:1022597616872

17. Gaugler JE et al (2009) Predictors of nursing home admission for persons with dementia. Med Care 47(2):191–198. https://doi.org/10.1097/MLR.0b013e31818457ce
18. Geda YE et al (2013) Neuropsychiatric symptoms in Alzheimer's disease: past progress and anticipation of the future. Alzheimer's Dement: J Alzheimer's Assoc 9(5):602–608. https://doi.org/10.1016/j.jalz.2012.12.001
19. Guaita A, Jones M (2011) A "prosthetic" approach for individuals with dementia? JAMA—J Am Med Assoc 305(4). https://doi.org/10.1001/jama.2011.28
20. Gustavsson A et al (2011) Cost of disorders of the brain in Europe 2010. Eur Neuropsychopharmacol 21:718–779
21. Haegerstrom-Portnoy G (2005) The Glenn A. Fry award lecture 2003: vision in elders—Summary of findings of the SKI study. Optom Vis Sci. https://doi.org/10.1097/01.OPX.000 0153162.05903.4C
22. Kavcic V et al (2006) Neurophysiological and perceptual correlates of navigational impairment in Alzheimer's disease. Brain. https://doi.org/10.1093/brain/awh727
23. Lyketsos CG et al (2012) Neuropsychiatric symptoms in Alzheimer's disease. Alzheimers Dement 7(5):532–539. https://doi.org/10.1016/j.jalz.2011.05.2410.Neuropsychiatric
24. Maluf A et al (2020) Structure and agency attributes of residents' use of dining space during mealtimes in care homes for older people. Health Soc Care Commun. https://doi.org/10.1111/hsc.13023
25. Mapstone M, Steffenella TM, Duffy CJ (2003) A visuospatial variant of mild cognitive impairment: getting lost between aging and AD. Neurology. https://doi.org/10.1212/01.WNL.000004 9471.76799.DE
26. Marquardt G (2011) Wayfinding for people with dementia: a review of the role of architectural design. Health Environ Res Des J. https://doi.org/10.1177/193758671100400207
27. Marquardt G, Bueter K, Motzek T (2014) Impact of the design of the built environment on people with dementia: an evidence-based review. HERD: Health Environ Res Des J. https://doi.org/10.1177/193758671400800111
28. Marquardt G, Schmieg P (2009) Dementia-friendly architecture: environments that facilitate wayfinding in nursing homes. Am J Alzheimer's Dis Other Dement. https://doi.org/10.1177/1533317509334959
29. McMonagle P et al (2006) The cognitive profile of posterior cortical atrophy. Neurology. https://doi.org/10.1212/01.wnl.0000196477.78548.db
30. Mendez MF et al (1990) Complex visual disturbances in Alzheimer's disease. Neurology. https://doi.org/10.1212/WNL.40.3_Part_1.439
31. Mendola JD et al (1995) Prevalence of visual deficits in Alzheimer's disease. Optom Vis Sci. https://doi.org/10.1097/00006324-199503000-00003
32. Jones M (1999) Gentlecare: changing the experience of Alzheimer's disease in a positive way. Hartley & Marks Publishers Inc., Vancouver, BC
33. Neargarder SA et al (2003) The impact of acuity on performance of four clinical measures of contrast sensitivity in Alzheimer's disease. J Gerontol Ser B Psychol Sci Soc Sci. https://doi.org/10.1093/geronb/58.1.P54
34. Olsson A et al (2012) My, your and our needs for safety and security: relatives' reflections on using information and communication technology in dementia care. Scand J Caring Sci. https://doi.org/10.1111/j.1471-6712.2011.00916.x
35. Rizzo M et al (2001) Simulated car crashes at intersections in drivers with Alzheimer disease. Alzheimer Dis Assoc Disord. https://doi.org/10.1097/00002093-200101000-00002
36. Salamone G et al (2009) Color discrimination performance in patients with alzheimer's disease. Dement Geriatr Cognit Disord. https://doi.org/10.1159/000218366
37. Shukla M et al (2017) Mechanisms of melatonin in alleviating Alzheimer's disease. Curr Neuropharmacol. https://doi.org/10.2174/1570159X15666170313123454
38. Uwajeh PC, Iyendo TO, Polay M (2019) Therapeutic gardens as a design approach for optimising the healing environment of patients with Alzheimer's disease and other dementias: a narrative review. Explore. https://doi.org/10.1016/j.explore.2019.05.002

39. West CG et al (2002) Is vision function related to physical functional ability in older adults? J Am Geriatr Soc. https://doi.org/10.1046/j.1532-5415.2002.50019.x
40. West SK et al (2002) How does visual impairment affect performance on tasks of everyday life? The SEE Project. Salisbury eye evaluation. Arch Ophthalmol. https://doi.org/10.1097/001 32578-200210000-00023
41. Whear R et al (2014) What Is the impact of using outdoor spaces such as gardens on the physical and mental well-being of those with dementia? A systematic review of quantitative and qualitative evidence. J Am Med Dir Assoc. https://doi.org/10.1016/j.jamda.2014.05.013
42. Wimo A et al (2013) The worldwide economic impact of dementia 2010. Alzheimer's Dement: J Alzheimer's Assoc 9:1–11
43. Wu YT et al (2015) Dementia in western Europe: epidemiological evidence and implications for policy making. Lancet Neurol. https://doi.org/10.1016/S1474-4422(15)00092-7

Light, Circadian Rhythms and Health

Roberto Manfredini, Rosaria Cappadona, Ruana Tiseo, Isabella Bagnaresi, and Fabio Fabbian

1 Chronobiology and Circadian Rhythms

Chronobiology is the branch of biomedical sciences that studies biological rhythms. Biological rhythms exist at any level of living organisms and, according to their cycle length, rhythms are classified in: (i) circadian (from the Latin 'circa-diem', with period of approximately 24 h), (ii) ultradian (period < 24 h, e.g., hours or minutes), and (iii) infradian (period > 24 h, e.g., weeks, months, seasons) [1]. Circadian rhythms are the most widely studied, since life on Earth is adapted to its rotation schedule, that is around 24 h. The periodic succession of night and day influences all living organisms, and the presence or absence of light influences the functioning of the biological clock. In fact, organisms have evolved internal circadian clocks capable of adapting to match precise cyclical 24 h fluctuations in the environment (e.g. light, food availability, predation) [2]. For nonhuman animals, restricting activities to the appropriate time is crucial to fitness and survival; for humans, temporal organization of physiology plays a pivotal role for both health and wellness [3]. Among nonhuman organisms, Cyanobacterias are the oldest known living system in the world (2.8 billion years) and were the first microbe to produce oxygen via photosynthesis, so

R. Manfredini (✉) · R. Cappadona · R. Tiseo · I. Bagnaresi · F. Fabbian
Dipartimento di Scienze Mediche, Università Di Ferrara, Ferrara, Italy
e-mail: roberto.manfredini@unife.it

R. Cappadona
e-mail: rosaria.cappadona@unife.it

R. Tiseo
e-mail: ruana.tiseo@unife.it

I. Bagnaresi
e-mail: clinicamedica@unife.it

F. Fabbian
e-mail: fabio.fabbian@unife.it

participating to the conversion of Earth's atmosphere. Cyanobacteria are considered the simplest organisms and the only prokaryotes in possess of a circadian clock, with physiological features similar to those of eukaryotes [4]. The dinoflagellate Gonyaulax polyedra (1.8 billion years) is characterized by cyclic organization with three circadian rhythms, one each for photosynthetic capacity, bioluminescence, and cellular division. The presence of a circadian oscillator is beneficial for a cell for at least two reasons: (i) no need to be restarted every day, even without a time cue; (ii) additional level of flexibility of response to a changing environment [5].

1.1 Molecular Organization

At the top of the hierarchical pyramid, there is the Suprachiasmatic Nuclei (SCN), comprised of approximately 30,000 neurons and located in the anterior hypothalamic region of the brain [6]. The SCN primarily coordinates the oscillator systems that regulate physiology and behavior. Its prominent role in animals has been demonstrated by ablation and transplantation studies [7]. In response to light exposure, the main molecular machinery is represented by a complex system of positive and negative molecular feedback loops that regulate the rhythmic expression of clock-controlled genes [8].

1.2 Circadian Individual Preference (Chronotype)

Horne and Ostberg first described the possibility of individual differences in circadian attitudes [9]. They ideated a self-assessment Morningness-Eveningness Questionnaire (MEQ), based on 19 items with a definite score each, identifying five categories: definitely Evening-type (E-types), moderately E-types, neither-type or Intermediate (I-types), moderately Morning-types (M-types), and definitely M-types. Multiple genes participate in the determination of chronotype, and among general population, chronotype exhibits a Gaussian distribution, with I-types accounting for 80%, and M-types and E-types for 10% each [10]. Moreover, individual chronotype is not immutable, and varies with age: E-type is more frequent in younger subjects, and M-type in the advancing age [11]. Our group reviewed the evidence on the relationship between chronotype, gender, and general health, with respect to four general issues: (a) "General and cardiovascular", (b) "Psychological and psychopathological", (c) "Sleep and sleep-related", and (d) "School and school-related". Basically, E-type was associated with: (a) unhealthy and dietary habits, smoking and alcohol drinking in younger subjects, diabetes and metabolic syndrome in adults; (b) impulsivity, anger, anxiety and depression disorders, risk taking behavior; (c + d) later bedtime and wake-up time, irregular sleep–wake schedule, subjective poor sleep, lower school performance [12].

2 Light

2.1 Mechanisms for Vision: Image and Non-image Forming

Adaptation to ambient light is achieved by two basic mechanisms for vision: image forming (IF) and non-image forming (NIF). IF function is mediated by the eye, with its specialized optical structures, rods and cones, that project an image to the retina. The presence of NIF photoreception in the mammalian retina emerged in the late 90s. Mice lacking rods and cones were still capable of synchronizing their circadian rhythms to light–dark cycles, due to an undiscovered photopigment/photoreceptor responsible for photoentrainment of circadian rhythms [13]. Specific projection neurons of the retina (retinal ganglion cells, or RGCs), transmit signals to a diverse group of subcortical target structures in the brain, including the SCN, innervated by a small subset of RGCs expressing the photopigment melanopsin (intrinsically photosensitive RGCs, or ipRGCs) [14]. Melanopsin-associated system and rod/cone system are complementary functions [15]. Cone and rod photoreceptors of the outer retina capture the image and send the information via ipRGCs, and the melanopsin signaling system plays multiple roles, e.g., NIF visual functions, photoentrainment of circadian rhythms, sleep regulation, and suppression of melatonin. Many blind subjects with no light perception show a circadian desynchronization, since the clock-controlled rhythms progressively shift in accordance with the internal near-24 h periodicity. Thus, consequent poor sleep, daytime somnolence and nighttime insomnia severely affect not only personal health, but social, academic, and professional life as well (Non-24 Sleep–Wake Rhythm Disorder, or NSWRD) [16].

2.2 Natural Light, Artificial Light?

For billion years, the light/dark cycle determined by the sun has regulated the endogenous circadian rhythmicity in almost all life forms. As the Earth rotates around its own axis, the ambient light undergoes predictable changes in intensity and spectrum as a function of solar angle. During the day the sun provides bright light, so that the ambient light is 1,000,000–100,000,000 times brighter than at night. On a clear night, starlight provides an illuminance of about ~0.001 lx, while moonlight is about ~0.2 lx. Until the XIX century, lightening during the night was warranted by fire. However, light from a fire is very different compared to the one generated by an electric light, with a wavelength strongly orientated toward the red end of the spectrum, and has much less impact on circadian rhythms than electric light [17]. In the modern era, lighting have been endorsed to more economic incandescent light sources. However, electric light has some problems, since intensity and spectral content often are not adequate during the day for circadian resetting, and excessive during the night. There are a number of aspects of the impact of Light-At-Night (LAN) on human circadian rhythmicity: (i) blue light is most effective, red the least; (ii) there is a dose–response effect; (iii) there is an influence on night-time sensitivity by daily light exposure; (iv) inter-individual differences in sensitivity exist [18]. Moreover, the availability of electric light allowed man to work over the 24 h of a day, giving rise to a series of health problems defined '*shift work desynchronization*'. Exposition to excessive nighttime light alters the sleep–wake schedule, and the efforts to remain awake at night and sleep during the day lead to sleep deprivation, since daytime sleep is more fragmented and less restorative [19]. In 2009, the European Commission began to replace incandescent lamps with more energy-efficient technologies. The development of high brightness Light-Emitting Diodes (LEDs) provided advantage for general illumination requiring smaller size screen, so that LEDs rapidly became the dominant technology for smartphones, tablets, iPads, e-readers, and LCD television sets [19].

2.3   *Light and Damage to the Retina*

Light can induce damage via three mechanisms: photomechanical, photothermal, and photochemical: (i) photomechanical, due to rapid increase in the amount of energy captured by the retinal pigment epithelium (RPE), with irreversible damage to the RPE and photoreceptors; (ii) photothermal, secondary to exposition to brief (100 ms–10 s) but intense light, capable to induce significant increase in the temperature of the retina and the RPE; and (iii) photochemical, following an exposition to light of high intensity in the visible range [19]. The exposure to blue light in the range of 400–470 nm may damage photoreceptors and RPE cells [20], and favors the development of age-related macular degeneration (AMD) or other retinal pathologies [21]. Moreover, studies on mice showed that light-induced damage to photoreceptors may vary by time of the day, since exposure to blue light during the night has more negative effects compared to the same exposure during the daytime [19].

2.4   *Light and Sleep*

Although the functions of sleep are not fully known, we spend roughly one third of our life sleeping. Thus, sleep certainly plays a crucial role for normal physiology and behavior. Sleep need is determined by many factors, including biological, environmental and social factors, and sleep habits have drastically changed through the course of human history. In particular, sleep timing changed from pre-industrial to industrial, and from rural to urban lifestyles, in great part due to the access to electricity, that allowed continuous working also in enclosed buildings [22]. The Munich ChronoType study showed that sleep on workdays shortened by 3.7 min/year over the past decade, while only that on work-free days remained the same [23]. Data based on satellite images of night-time illumination (US Defense Meteorological

Satellite Program), combined with country-level data from the World Health Organization, showed that exposure to artificial LAN can explain about 70% of the observed variation of overweight and obesity prevalence rates among females and males in more than eighty countries worldwide [24]. Even one only hour of sleep deprivation can favor disruption of cardiovascular health, and a higher incidence of myocardial infarctions have been observed following the Daytime Saving Time spring shift [25, 26].

2.5 Urban Light

Since the 70s, the Defense Meteorological Satellite Program's Operational Linescan System records nighttime light data on earth. In a nonurban environment, away from artificial lights, nocturnal light levels range from $\approx 1 \times 10^{-4}$ lx in typical moonlight to 0.1–0.3 lx during the week around the full moon [27]. The artificial light of a common streetlight is ≈ 5 lx and a parking lot light in a shopping mall is ≈ 20 lx. A study on more than 19,000 adult subjects living in 15 US states, showed that living in areas with greater Outdoor Nighttime Lights (ONL) was significantly associated with delayed bedtime and wake up time, shorter sleep duration, and increased daytime sleepiness, dissatisfaction with sleep quantity and quality, and a diagnostic profile congruent with a circadian rhythm disorder [28]. Although nighttime lights in our streets and cities improve the overall safety of people and traffic, they are associated with modifications in human sleep behaviors and individual circadian rhythms. The scale of exposure to artificial lighting is increasing as cities switch streetlamps to LEDs. In the United States, 10% of all street lighting has been converted, and the city of Milan, in Italy, was the first in Europe to do so on large scale [29].

2.6 Domestic Light

Smartphones are now ubiquitous in daily life, and their average use time has dramatically increased. Every three out of four children or adolescents have screen-based devices in their bedrooms, and 60% of them view or interact with screens in the hour before bedtime [30]. The type of light influences the magnitude of melatonin suppression. Light of short wavelength has a larger suppressing effect compared to light with longer wavelengths [31] and exposure to blue light LED of self-luminous tablets significantly reduces melatonin levels, compared to orange LED [32]. Moreover, the magnitude of suppression by a light stimulus in children is almost twice that of adults [33]. Also the use of electronic books on a light-emitting device (LE-eBook) is increasing. Compared with reading a print-book, subjects reading a LE-eBook showed, among others, suppressed evening levels of melatonin and delay to fall asleep [34]. Screen time and media use in school-age youth and teenagers was adversely associated with sleep health [35]. Potential mechanisms include: (a) time displacement; (b) psychological stimulation based on media content; and (c) the effects on circadian timing, sleep physiology, and alertness [36]. The widespread use of electronic devices during nighttime exposes the youngest generation to a transformation into E-type, and individual chronotype significantly impacts on sleep and sleep problems, associated with daytime tiredness, poor school performance and psychological problems. The American Academy of Pediatrics published specific policy statements about media use in children and adolescents [37].

3 Beneficial Effects of Light Modulation in Special Settings

3.1 Intensive Care Units

Intensive care unit patients are exposed to pathologic wakefulness, poor quality of daytime sleep, nocturnal sleep fragmentation, and altered sleep patterns (absence of slow wave and REM phase) [38]. Recent studies confirmed the effectiveness of simple measures to reduce sound/light exposure as also an important Evidence-Based Design (EBD) strategy [39, 40].

3.2 Elderly Care Homes

In elderly subjects, some circadian rhythm disorders, e.g., disruption of sleep–wake cycle, could be explained by either impairment of melanopsin RGC and decrease of the spectrum of blue light transmission with age [41, 42]. Moreover, since exposure to blue light increases alertness and stimulate cognitive functions, the lack of blue light during daytime can explain sleep problems and cognitive decline [43]. These aspects are particularly evident for older people living in elderly care homes, where environmental (outdoor and indoor light) and age-related physiological factors may favor a light-deprived environment. Hopkins et al. investigated the effect of increasing indoor light levels with blue-enriched white lighting compared with control white lighting, on rest-activity rhythms, performance, self-reported mood, sleep, and alertness in older residents [44]. Blue-enriched light produced either positive effects (increased daytime activity, reduced anxiety), and negative ones (increased nighttime activity, reduced sleep efficiency and quality).

3.3 Psychiatric Settings

Hospital psychiatric units often maintain lighting during nighttime, to facilitate the observation of patients, but this may contribute to disrupt sleep. Martin et al. investigated in an old age psychiatry inpatient setting the effects of changing night-lights from broad-band white to narrow-band red on some items, such as amount of sleep observed, "as required" medication administered and number of falls [45]. They reported more observations of sleep and fewer 'as required' medication administrations during red light than white light period, as well as better sleep and minor agitation at night.

3.4 Workplaces

Adequate illumination is important also in workplaces. A study on control room staff members of petrochemical industry showed that the use of blue-enriched white light illumination, compared to normal lighting conditions, reduced sleepiness and melatonin rhythm, and gave ergonomic advantage, by decreasing working memory errors, omission errors and reaction time during sustained attention task [46].

4 Natural Light and Green Spaces: The Healing Gardens

A growing attention is now given to urban planning and public health. The introduction of the social model of health has stressed the importance of the determinants of health such as socioeconomic, cultural, and environmental conditions, in addition to living and working conditions [47]. However, striking contradictory behaviors still exist. On one hand, the Decalogue for Healthy Cities (Erice 50 Charter "Strategies for Diseases Prevention and Health Promotion in Urban Areas"), aimed to design effective strategic actions and best practices to develop urban regeneration interventions and improve the urban quality of contemporary cities, was unanimously approved [48]. On the other hand, a law of Lombardy Region of Italy (March 2017) allowed to live in basements (defined as buildings partly below curb level but with at least one-half of its height above the curb), despite the available evidence of exposition of residents to an increased risk of cardio-respiratory diseases and neoplasms [49, 50].

There is a rapidly growing body of research on the relationship between green space and mental and physical well-being. Green space is defined as "an area of grass, trees, or other vegetation set apart for recreational or aesthetic purposes in an otherwise urban environment" [51]. Thus, a careful attention has been given to concept of garden, proposed as a place of care, a promoter of restoration of the human being. Mechanisms leading stress regulation, level of attention and organization, focus and fascination, are recognized at the origin of restoration processes.

Quoting with Pringuey-Criou '*the garden opens the door to our interiority and prepares the interpersonal meeting*' [52]. The healing benefits of gardens have been object of different strategies worldwide, including open community gardens affiliated with healthcare institutions, or designed outdoor spaces in healthcare contexts, such as rooftop hospital gardens, inspired from rooftop gardens benefits [53–55]. According to a health care design focused on "patient-centered care" and "healing environments" to alleviate the stressful nature of serious illness, health care facilities deserve particular attention. On one hand, they may include resource center, where patients and families can educate themselves about their illness, or space in rooms to allow family members to remain with patients. On the other, considerable attention has to be addressed to ambient features, such as adequate lighting, water features, green spaces, and healing gardens. All of these aspects can produce positive effects for patients and employees [55]. Thus, a greater integration is needed between the different professionals involved in urban planning and in health care analysis. This is crucial to identify appropriate research on creation of green areas, capable to conciliate environmental sustainability and health requirements [56]. Here some examples, with reference to special groups of patients.

4.1 Patients with Alzheimer Disease

Maintaining quality of life is important in dementia care, and sensory gardens and horticultural activities are increasingly used also in these patients. The French National Alzheimer Plan 2008–2012, based on the implementation of units specialized in cognitive rehabilitation and psycho-behavioral therapy of Alzheimer's disease patients, recognized healing gardens as integral part of these Units [57]. A systematic search on the effects of gardens reported findings mainly on issues related to behavior, affect and well-being. Sleep pattern, well-being and functional level improved, as well as the use of psychotropic drugs, incidents of serious falls, sleep and sleep pattern [58].

4.2 People Living in Elderly Care Homes

Older people in residential facilities often suffer from complex health problems, such as multi-morbidity, reduced cognitive capacity, inability to perform activities of daily living, depression, anxiety, and pain. Beneficial relationship between gardens in residential facilities and resident health may include contact with nature, air quality, physical activity, social cohesion and stress reduction. A study based on questionnaires administered to residents at Swedish facilities for older people, reported positive significant effects on self-perceived health. Garden greenery enhanced sense of being away, afforded possibilities to experience the outdoor environment as interesting, and encouraged visitation [59].

4.3 *Children*

Children are spending more time indoors while pediatric mental and behavioral health problems are increasing. Access to green space is associated with improved mental well-being, overall health and cognitive development [60].

5 Conclusions

Healthy lighting design represents a novel important ethical issue. A global campaign in Japan to increase public awareness about the appropriate use of blue-rich LEDs and to protect the youngest population from the potentially harmful impact of LED screens was addressed to school teachers and parents, instructed on the possible effect of smartphones, tablets, computers; and to building industry and architects, called to maximize the availability of natural daylight during daytime hours [61]. The American Medical Association (AMA) issued a warning report entitled "Light Pollution: Adverse Health Effects of Nighttime Lighting" [62], recognizing that *"exposure to excessive light at night, including extended use of various electronic media, can disrupt sleep or exacerbate sleep disorders, especially in children and adolescents."*, and stating that *"disruption of circadian rhythmicity and sleep from the indiscriminate use of electric light at night may well increase risk of many of the diseases of modern life."* Again, the AMA issued the policy statement "Guidance to Reduce Harm from High Intensity Street Lights" to help in the selection from the different options of LED lighting. Examples of good lighting strategies exist. The Van Gogh village (Nuenen, the Netherlands), applied a method to lower street lights by 80% during low-activity times and switch them up when a pedestrian, cyclist or car approaches. After an expensive initial investment for installation, a further significant payback was obtained, since energy and maintenance costs reduced by 62% [30]. The architect and lighting designer Karolina Zielinska-Dabkowska suggested to policymakers a series of interventions, i.e., better use of natural light indoors during the day, rewarding policy for building practices and technologies that harness natural light, use of sustainable night-time illumination guidelines in the municipal urban lighting plans, attention to lighting of street and security, for walking, cycling and driving [30]. Light has always been a synonymous of happiness and positivity. Now we know that a correct light/dark alternation is crucial for respect of endogenous circadian rhythms and health. Each of us, at any level, may give a personal contribution. Let's remember it, every day.

References

1. Manfredini R, Boari B, Salmi R, Malagoni AM, Manfredini F (2007). Circadian rhythm effects on cardiovascular and other stress-related events. In: George Fink, Encyclopedia of Stress, vol 1, 2nd edn. Academic Press, Oxford, pp 500–505

2. West AC, Smith L, Ray DW, Loudon ASI, Brown TM, Bechtold DA (2017) Misalignment with the external light environment drives metabolic and cardiac dysfunction. Nat Commun 8:417

3. Bedrosian TA, Nelson RJ (2017) Timing of light exposure affects mood and brain circuits. Transl Psychiatry 7:e1017

4. Cohen SE, Golden SS (2015) Circadian rhythms in cyanobacteria. Microbiol Mol Biol Rev 79:373–385

5. Dunlap JC, Loros JJ (2017) Making time: conservation of biological clocks from fungi to animals. Microbiol Spectr 5:3

6. Mohawk JA, Takahashi JS (2011) Cell autonomy and synchrony of suprachiasmatic nucleus circadian oscillators. Trends Neurosci 34:349–358

7. Ralph MR, Foster RG, Davis FC, Menaker M (1990) Transplanted suprachiasmatic nucleus determines circadian period. Science 247:975–978

8. Crnko S, Cour M, Van Laake LW, Lecour S (2018) Vasculature on the clock: circadian rhythm and vascular dysfunction. Vascul Pharmacol 108:1–7

9. Horne JA, Ostberg O (1976) A self-assessment questionnaire to determine morningness-eveningness in human circadian rhythms. Int J Chronobiol 4:97–110

10. Ashkenazi IE, Reinberg AE, Motohashi Y (1997) Interindividual differences in the flexibility of human temporal organization: pertinence to jetlag and shiftwork. Chronobiol Int 14:99–113

11. Paine SJ, Gander PH, Travier N (2006) The epidemiology of morningness/eveningness: influence of age, gender, ethnicity, and socioeconomic factors in adults (30–49 years). J Biol Rhythms 21:68–76

12. Fabbian F, Zucchi B, De Giorgi A et al (2016) Chronotype, gender and general health. Chronobiol Int 33:863–882

13. Freedman MS, Lucas RJ, Soni B et al (1999) Regulation of mammalian circadian behavior by non-rod, non-cone, ocular photoreceptors. Science 284:502–504

14. Dhande OS, Stafford BK, Lim JA, Huberman AD (2015) Contributions of retinal ganglion cells to subcortical visual processing and behaviors. Annu Rev Vis Sci 1:291–328

15. Paul KN, Saafir TN, Tosini G (2009) The role of retinal photoreceptors in the regulation of circadian rhythms. Rev Endocr Metab Disord 10:271–278

16. Quera Salva MA, Hartley S, Léger D, Dauvilliers YA (2017) Non-24 h sleep-wake rhythm disorder in the totally blind: diagnosis and management. Front Neurol 8:686

17. Stevens RG, Zhu Y (2015). Electric light, particularly at night, disrupts human circadian rhythmicity: is that a problem? Philos Trans R Soc Lond B Biol Sci 370:1667

18. LeGates TA, Fernandez DC, Hattar S (2014) Light as a central modulator of circadian rhythms, sleep and affect. Nat Rev Neurosci 15:443–454

19. Tosini G, Ferguson I, Tsubota K (2016) Effects of blue light on the circadian system and eye physiology. Mol Vis 22:61–72

20. Kuse Y, Ogawa K, Tsuruma K, Shimazawa M, Hara H (2014) Damage of photoreceptor-derived cells in culture induced by light emitting diode-derived blue light. Sci Rep 4:5223

21. Klein R, Klein BEK, Jensen SC, Cruickshanks KJ (1998) The relationship of ocular factors to the incidence and progression of age-related maculopathy. Arch Ophthalmol 116:506–513

22. Bin YS, Marshall NS, Glozier N (2012) Secular trends in adult sleep duration: systematic review. Sleep Med Rev 16:223–230

23. Roenneberg T, Allebrandt KV, Merrow M, Vetter C (2012) Social jetlag and obesity. Curr Biol 22:939–943

24. Rybnikova NA, Haim A, Portnov BA (2016) Does artificial light-at-night (ALAN) exposure contribute to the worldwide obesity pandemic? Int J Obes (London) 40:815–823

25. Manfredini R, Cappadona R, Modesti PA, Fabbian F (2018) Daylight saving time and cardiovascular health. Int Emerg Med 13:641–646

26. Manfredini R, Fabbian F, De Giorgi A et al (2018) Daylight saving time and myocardial infarction: should be we worried? A review of the evidence. Eur Rev Med Pharmacol Sci 22:750–755
27. Falchi F, Cinzano P, Elvidge CD, Keith DM, Haim A (2011) Limiting the impact of light pollution on human health, environment and stellar visibility. J Environ Manag 92:2714–2722
28. Ohayon MM, Milesi C (2016) Artificial outdoor nighttime lights associate with altered sleep behavior in the American general population. Sleep 39:1311–1320
29. Zielinska-Dabkowska K (2018) Make lighting healthier. Nature 553:274–276
30. Hysing M, Pallesen S, Stormark KM, Jakobsen R, Lundervold AJ, Sivertsen B (2015) Sleep and use of electronic devices in adolescence: results from a large population-based study. BMJ Open 5:e006748
31. Cajochen C, Frey S, Anders D et al (2011) Evening exposure to a light-emitting diodes (LED)-backlit computer screen affects circadian physiology and cognitive performance. J Appl Physiol 110:1432–1438
32. Wood B, Rea MS, Plitnick B, Figueiro MG (2013) Light level and duration of exposure determine the impact of self-luminous tablets on melatonin suppression. Appl Ergon 2013(44):237–240
33. Higuchi S, Nagafuchi Y, Lee SI, Harada T (2014) Influence of light at night on melatonin suppression in children. J Clin Endocrinol Metab 2014(99):3298–3303
34. Changa AM, Aeschbacha D, Jeanne F, Duffya JF, Czeisler CA (2015) Evening use of light-emitting eReaders negatively affects sleep, circadian timing, and next-morning alertness. Proc Natl Acad Sci 112:1232–1237
35. LeBourgeois MK, Hale L, Chang AM et al (2017) Digital media and sleep in childhood and adolescence. Pediatrics 140:S92–S96
36. Chassiakos RY, Radesky J, Christakis D et al (2016). AAP Council on Communications and Media. Child Adolesc Digit Med Pediatr 138:e20162593
37. American Academy of Pediatrics (2016) AAP Council on Communications and Media. Media School-Aged Child Adolesc Pediatr 138:e20162592
38. Korompeli A, Muurlink O, Kavrochorianou N, Katsoulas T, Fildissis G, Baltopoulos G (2017) Circadian disruption of ICU patients: a review of pathways, expression, and interventions. J Crit Care 38:269–277
39. Bion V, Lowe AS, Puthucheary Z, Montgomery H (2018) Reducing sound and light exposure to improve sleep on the adult intensive care unit: an inclusive narrative review. J Intensive Care Soc 19:138–146
40. Brambilla A, Rebecchi A, Capolongo S (2019). Evidence based hospital design. A literature review of the recent publications about the EBD impact of built environment on hospital occupants' and organizational outcomes. Ann Ig 31:165–180
41. Esquiva G, Lax P, Pérez-Santonja JJ, García-Fernández JM, Cuenca N (2017) Loss of melanopsin-expressing ganglion cell subtypes and dendritic degeneration in the aging human retina. Front Aging Neurosci 9:79
42. Daneault V, Hébert M, Albouy G et al (2014) Aging reduces the stimulating effect of blue light on cognitive brain functions. Sleep 37:85–96
43. Lockley SW, Gooley JJ (2006) Circadian photoreception: spotlight on the brain. Curr Biol 16:R795–R797
44. Hopkins S, Morgan PL, Schlangen LJM, Williams P, Skene DJ, Middleton B (2017) Blue-enriched lighting for older people living in care homes: effect on activity, actigraphic sleep, mood and alertness. Curr Alzheimer Res 14:1053–1062
45. Martin D, Hurlbert A, Cousins DA (2018) Sleep disturbance and the change from white to red lighting at night on old age psychiatry wards: a quality improvement project. Arch Psychiatr Nurs 32:379–383
46. Motamedzadeh M, Golmohammadi R, Kazemi R, Heidarimoghadam R (2017) The effect of blue-enriched white light on cognitive performances and sleepiness of night-shift workers: a field study. Physiol Behav 177:208–214

47. Capolongo S, Lemaire N, Oppio A, Buffoli M, Roue Le Gall A (2016) Action planning for healthy cities: the role of multi-criteria analysis, developed in Italy and France, for assessing health performances in land-use plans and urban development projects. Epidemiol Prev 40:257–264

48. D'Alessandro D, Arletti S, Azara A et al (2017) Strategies for disease prevention and health promotion in urban areas: the Erice 50 Charter. Ann Ig 29:481–493

49. Mezzoiuso AG, Gola M, Rebecchi A et al (2017) Indoors and health: results of a systematic literature review assessing the potential health effects of living in basements. Acta Biomed 88:375–382

50. Gola M, Signorelli C, Buffoli M, Rebecchi A, Capolongo S (2017) Local health rules and building regulations: a survey on local hygiene and building regulations in italian municiples. Ann Ist Super Sanità 53:223–230

51. Oxford University Press. Green space. From Oxford Living Dictionaries (2017). https://en.oxf orddictionaries.com/definition/green_space

52. Pringuey-Criou F (2015) Healing garden: primary concept. Encephale 41:454–459

53. George DR, Rovniak LS, Kraschnewski JL, Hanson R, Sciamanna CN (2015) A growing opportunity: community gardens affiliated with US hospitals and academic health centers. Prev Med Rep 2:35–39

54. Buffoli M et al (2018) Green SOAP. A calculation model for improving outdoor air quality in urban contexts and evaluating the benefits to the population's health status. In: Mondini G et al (eds) Integrated evaluation for the management of contemporary cities. Springer, Green Energy and Technology, pp 453–467

55. Mroczek J, Mikitarian G, Vieira EK, Rotarius T (2005) Hospital design and staff perceptions: an exploratory analysis. Health Care Manag (Frederick) 24:233–244

56. D'Alessandro D et al (2015) Green areas and public health: improving wellbeing and physical activity in the urban context. Epidemiol Prev 39(4 Suppl 1):8–13

57. Rivasseau Jonveaux T, Batt M, Fescharek R et al (2013) Healing gardens and cognitive behavioral units in the management of Alzheimer's disease patients: the Nancy experience. J Alzheimers Dis 34:325–338

58. Gonzalez MT, Kirkevold M (2014) Benefits of sensory garden and horticultural activities in dementia care: a modified scoping review. J Clin Nurs 23:2698–2715

59. Dahlkvist E, Hartig T, Nilsson A, Högberg H, Skovdahl K, Engström M (2016) Garden greenery and the health of older people in residential care facilities: a multi-level cross-sectional study. J Adv Nurs 72:2065–2076

60. McCormick R (2017) Does access to green space impact the mental well-being of children: a systematic review. J Pediatr Nurs 37:3–7

61. Hatori M, Gronfier C, Van Gelder RN et al (2017) Global rise of potential health hazards caused by blue light-induced circadian disruption in modern aging societies. NPJ Aging Mech Dis 3:9

62. Stevens RG, Brainard GC, Blask DE, Lockley SW, Motta ME (2013) Adverse health effects of nighttime lighting: comments on American Medical Association policy statement. Am J Prev Med 45:343–346

Green Spaces and Public Health in Urban Contexts

Andrea Lauria

1 Health Determinants, Physical Activity and Green Areas

Individuals health determinants and all the aspects that impact on the individual health status can be clustered into four main categories. The first one is related to non-modifiable factors such as age, sex, genetics, etc. The second category involves all the socio-economic factors such as level of education, income, social support networks, etc. The third is lifestyle which means physical activity, nutrition, substance use, etc. Finally, the last but not the least important category embeds all the aspects related to the environment where the individuals live.

As widely stated and reported by the World Health Organization (WHO), environmental conditions have an important role in impacting and influencing well-being and in reducing the population's health risk. For example, air quality and pollution exposure are perhaps the best-known and most studied aspects, but there is now strong evidence supporting the importance of urban green areas and their role in the prevention and contrast of some diseases, especially heart related ones [1, 2]. Considering that in Europe cardiac and cardiovascular pathologies (so-called chronic degenerative diseases) are a big threat, with over four million deaths annually, the impact of such green areas is very relevant [3].

Those aspects are very important, and the Public Health task is to involve more and more people into healthy habits. Constant research and awareness are therefore necessary to engage people in having active lifestyle, for example through the improvement of cycling activities in urban areas [4].

Addressing this public health problem in a risk reduction perspective, emerged the need to focus on the prevention aspect that comes from having an active lifestyle. In fact, it has been observed a 30–50% reduction in the relative risk of coronary heart disease among the physically active population which exercises moderately every or

A. Lauria (✉)
Regione Veneto, AULSS 9 Scaligera Verona, Verona, Italy
e-mail: andrea.lauria@aulss9.veneto.it

© The Author(s), under exclusive license to Springer Nature Switzerland AG 2023
S. Capolongo et al. (eds.), *Therapeutic Landscape Design*,
PoliMI SpringerBriefs, https://doi.org/10.1007/978-3-031-09439-2_9

almost every day, compared to the sedentary ones, considering all other risk factors being equal [5]. In Public Health terms, the importance of concrete policies, strategies and tools encouraging an increase in physical activity is reasonably important and, for example, the implementation of urban green areas can play a consistent role. It is therefore evident that when physical strategies, such as urban design, are capable to impact lifestyle, they become a health determinant factor to be carefully considered [6]. Indeed, having different types of well-equipped green spaces, gardens or parks in an urban context is a great opportunity for all social classes to be physically active in outdoor areas. Furthermore, urban green planning can also become part of a network that connects urban areas with the countryside and natural parks, establishing a "green system" which we can act as a *plan* able to influence high-level general local planning rules and guidelines [7].

It can ultimately be said that a city with a large number of usable and accessible green spaces can give residents the opportunity (at no cost) to increase exercise and outdoor sports and, eventually, to reduce cardiovascular risk.

2   Green Spaces and Health: Insights from the Literature

Among the most interesting and recent international studies, two of them are hereafter discussed as they reveal a correlation between green spaces and health. In a 2002 Japanese study the authors examined the association between the presence of green areas near the homes of 3,144 elderly Tokyo residents and their survival rates [8]. The study shown that five-year survival rates were directly proportional to the space available for walking, the number of parks and the tree-lined streets near people's homes as well as to the number of hours of sunlight the homes were exposed to and the willingness to continue living in the same area for the existence of a social network. Several thousand of elderly people were interviewed in the study in order to understand whether urban structure influenced also life expectancy. The research found that sedentary lifestyle is a central risk factor for morbidity, premature mortality and for the reduction of physical functions in the older population.

Nevertheless, a careful approach is needed when verifying results: an association between health outcomes and living environments is controversial and difficult to demonstrate as the impact on health often only becomes visible many years later, and because many other physical, social and economic factors influence both the environmental conditions of residential areas and their citizens' health status. The authors, through a longitudinal cohort study and considering multiple variables related to physical, economic and social realities analysed the independent effect of green spaces on the longevity of older citizens; they also checked whether these spaces constitute an environment which can support the promotion of senior citizens' health in densely populated urban areas. In summary, the five-year survival probability was directly proportional to: the space available for walking ($p < 0.01$); the number of parks and tree-lined roads near the home ($p < 0.05$); the hours of sun exposure (p < 0.01) and the desire to continue living in the same neighbourhood (p < 0.01).

The second research is a 2009 study of 350,000 patients by 195 Dutch General Practitioners (GP) showed that for 15 of the 24 diseases examined the frequency of chronic diseases (coronary heart disease, angina, heart attack, skeletal disorders, anxiety, depression, respiratory infections, headache, dizziness, urinary tract infections and diabetes) was lower in those living less than 1 km away from parks or green areas [9]. This second study, carried out on the Dutch population, diverges from the Japanese study mainly because it considers a large part of the population, verifying data objectively treated without the use of an individual questionnaire. The cartographic platform used highlight a sort of zoning with a map of 39 different environmental territory uses within a grid with a base module of 2.5×2.5 km. The green spaces represented occupy at least 40% of the cell. It was possible to geo-reference citizens via postcodes and examine places of residence in relation to the presence of adequate green spaces. Furthermore, all the correlations between green spaces and physical or mental health were highlighted.

3   Health Inequalities and Access to Green Spaces

These two studies open the debate on the presence or absence of green areas within urban context and their influence on a population's health conditions. This concept can also be looked at in terms of inequalities. The theme of inequality and health has been constantly studied for many years by one of the world's leading experts in the field, Professor Sir Michael Marmot [10]. The starting point is the interpretation of health inequalities as social injustice, directly related to inequalities in education, income, access to services and differences in quality of environment and life. Among environmental aspects, the presence of open spaces and urban green areas is one of the health determinants that affect all age groups and constitutes the fundamental resource for obtaining sustainable and healthy collective spaces. As shown in Fig. 1, there is a close relationship between the use of green spaces and the level of protection from cardiovascular diseases given to those from the less wealthy social classes [11]. It is interesting to note how subjects in the various groups with a higher income maintain a low mortality risk either in the presence or absence of green spaces. On the other hand, the presence or lack of green spaces plays a strategic role of protection with very different values among lower-income groups. As can be seen among the five groups of people analysed, the incidence of death from circulatory disease is reduced with the higher presence of green spaces.

In terms of population access to green spaces, the observations of the English Institute of Health Equity emphasise the importance of actions for the benefit of the less wealthy social classes [12]. During the period studied, from March 2012 to February 2013, there was a lower number of visits and use of green areas by the low-income population compared to the better-off social classes. It is therefore not only a question of facilities or accessibility, but also of setting up policies for

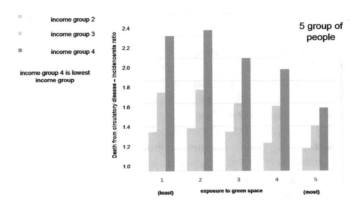

Fig. 1 Health benefits of exposure to green space [11]

the promotion and use of open spaces to all [13, 14]. It is indeed evident that the culturally and economically well-equipped citizens, who already have tools enabling them to quickly grasp and make the most beneficial life choices, view the presence of green spaces and paths as enriching and a sign of better-quality urban standards. It is therefore clear that the wider part of the population (including young people and the elderly) has to be targeted as those who often need a "gentle push" to make these choices.[1] To use a more theatrical term, the target are those who need to be presented with a suitable stage to start playing a different role from the one they have had so far.

4   Discussion: Strategies, Tools and the Italian Situation

It is not the intention of this brief discussion to identify a way to establish "inviting" green city spaces, but it is important to point out that the history of a city often does not include providing its inhabitants with suitable green spaces. Their existence in urban areas is constantly tied in with a city's history, reflecting morphological aspects and historical events, accompanying economic fortunes and misfortunes, and representing power structures and social relations.

We have inherited *gardens of princes* and *civil magnificence*, the parks and great avenues of the past that together with public and private neighbourhood areas make up the green spaces of our cities. What we have to face today is the reorganization of the network of green spaces to make them instruments of psychophysical health and social cohesion, as well as of urban quality.

---

[1] The term "gentle push" comes from "Nudge of the Book" by Richard Thaler and Cass Sustein, published in 2009 by Feltrinelli, who propound policies of "libertarian paternalism" to guide people's life choices of the population.

It is now very clear that the *standard*, the reference for the quantitative dimensioning of green areas is sterile: everything is reduced to an accounting of square meters, areas or volumes. The dissatisfaction of citizens and the difficulty in using green spaces does not derive mainly from quantitative insufficiency, as the verification of per capita availability of green spaces, but from the relationship between green spaces and the city, the quality of the green spaces, their accessibility and their functions.

Unfortunately, the themes of accessibility, facilities, form and figurative relationship with the rest of the city are often neglected, or even considered superfluous. Even rarer are examples of design attempting to integrate the different types of green into the structure of the city, conscious contributions to the quality of the environment and urban health, including any consequent psychological and social benefits. There is the growing need of multidisciplinary tools and instruments to evaluate and measure the qualities of different spaces also in terms of accessibility, especially when it comes to green areas [15, 16]. The absence of government regulations and supra-municipal green area management leads to difficulties in interpreting the role of green areas in terms of planning, through the adoption of a *Green Plan*. There are few attempts by municipal administrations to coordinate and manage this topic and to see all the roles that green areas can play.

The field is complex and there are many functions and aspects to be evaluated, such as the absorption of CO_2 and atmospheric pollutants, thermoregulation to reduce heat islands, the water cycle and more generally the improvement of soil hydrology, water purification and the production of food and raw materials, relevant because they have the potential to be widespread and short-chain: in other words, we can speak of "green infrastructure" [17].

To mention just some of the unfortunately few Italian experiences, the Green Plans of Milan and Reggio Emilia and the preliminary plans of Genoa should be noted. We can synthetically summarise these as evidence of the transition from a dimensional tool to a scenario dimension. The contributions made by regulations are also interesting. First of all, is important to mention the Turin experience which goes far beyond maintenance and organisational aspects and address citizen involvement in the management of green areas. Venice was among the first examples in Italy where the activation process involved citizens' committees, councillors, professional associations, universities, etc. defining regulations that called for citizens' co-responsibility and their involvement in planning and management activities. In terms of participation, the experience of the Bologna Urban Centre is an example of the ability and willingness to involve a city's social, cultural and economic resources. Finally, in the city of Bologna the topic of urban garden has been relaunched through different design competitions as a functional and social characteristic of the historical city.

Another interesting process that involves citizens and public administration in a social network are "subsidiarity pacts", agreements between citizens and the Municipality in which the common good, such as parks or green areas, can be regenerated for social and urban recovery purposes through a "take in charge" by citizens. At the moment, in Verona there are approximately 50 "subsidiarity pacts" between the Municipality and groups of citizens for the regeneration of green areas.

5   Conclusions

In conclusion it can reasonably be affirmed that the theme of open spaces, the role of urban green spaces and their links with the city are highly necessary and useful, and require new forms, even experimental ones, to become a regular feature of the urban landscape. New European policies are encouraging this theme, describing it as a network, an infrastructure and a system [18]. Nevertheless, the impulse to include the theme of health in *Green Plans* is still in the early research stages. In particular, a recent research, from 2019, published in The Lancet Planetary Health focuses attention on the protection of urban greenery from premature mortality through nine longitudinal studies involving seven countries, including Italy. A systematic study, a meta-analysis, which pooled the results, with strong evidence that the increase in urban greenery near homes is significantly associated with reduced premature mortality. This research, still in progress, should also produce indicators to increase green infrastructures in cities, which we have seen, certainly improve public health and mitigate the impacts of climate change, making cities better habitats for living. [19].

It is desirable that green areas should become a tool to promote active lifestyles, with participatory actions for the creation and management of spaces, which are a useful contribution to social cohesion.

References

1. WHO Regional Office for Europe. Urban Green Space Interventions and Health. A review of impacts and effectiveness (2017). https://www.cbd.int/health/who-euro-green-spaces-urbanh ealth.pdf
2. Buffoli M, Rebecchi A, Gola M, Favotto A, Procopio GP, Capolongo S (2018) Green soap. A calculation model for improving outdoor air quality in urban contexts and evaluating the benefits to the population's health status. In: Mondini G, Fattinnanzi E, Oppio A, Bottero M, Stanghellini S (eds) Integrated evaluation for the management of contemporary cities. Springer, Green Energy and Technology, pp 453–467. https://doi.org/10.1007/978-3-319-78271-3_36
3. Nichols M, Townsend N, Luengo-Fernandez R, Leal J, Gray A, Scarborough P, Rayner M (2012) European cardiovascular disease statistics 2012. European Heart Network, Brussels, European Society of Cardiology, Sophia Antipolis. ISBN 978-2-9537898-1-2. https://www.escardio.org/static_file/Escardio/Press-media/press-releases/2013/EU-cardiovascular-disease-statistics-2012.pdf
4. Rebecchi A, Boati L, Oppio A, Buffoli M, Capolongo S (2016) Measuring the expected increase in cycling in the city of Milan and evaluating the positive effects on the population's health status: a community-based urban planning experience. Ann Ig 28(6):381–391. https://doi.org/10.7416/ai.2016.2120
5. Ministry of Health. Clinical Evidence, Italian edition (2001)
6. Capolongo S, Rebecchi A, Dettori M, Appolloni L, Azara A, Buffoli M, Capasso L, Casuccio A, Conti Oliveri G, D'Amico A, Ferrante M, Moscato U, Oberti I, Paglione L, Restivo V, D'Alessandro D (2018) Healthy design and urban planning strategies, actions, and policy to achieve salutogenic cities. Int J Environ Res Public Health 15(12):2698. https://doi.org/10.3390/ijerph15122698

7. Vittadini MR, Bolla D, Barp A (Eds.) (2015) Spazi verdi da vivere: il verde fa bene alla salute. Il prato editore
8. Takano T (2002) Urban residential environments and senior citizens' longevity in megacity areas: the importance of walkable green spaces. J Epidemiol Community Health 56:913–918
9. Maas J, Verheij RA, de Vries S et al (2009) Morbidity is related to a green living environment. J Epidemiol Community Health 63:967–973. https://doi.org/10.1136/jech.2008.079038
10. Marmot M (2010) Fair society healthy lives, the Marmot review. Strategic review of health inequalities in England post-2010, UK. ISBN 978-0-9564870-0-1. www.ucl.ac.uk/marmot review
11. Mitchell R, Popham F (2008) Effect of exposure to natural environment on health inequalities: an observational population study. Lancet 372:1655–1660. https://doi.org/10.1016/S0140-673 6(08)61689-X
12. Hunt A, Burt J, Stewart D (2013) Monitor of Engagement with the Natural Environment: a pilot for an indicator of visits to the natural environment by children - interim findings from Year 1 (March 2013 to February 2014). Natural England Commissioned Reports, Number 166. ISBN 978-1-78354-152-2. www.gov.uk/natural-england
13. Mosca EI, Herssens J, Rebecchi A, Capolongo S (2019) Inspiring architects in the application of design for all: knowledge transfer methods and tools. J Access Design All 9(1):1–24. https://doi.org/10.17411/jacces.v9i1.147
14. Mosca EI, Capolongo S (2018) Towards a universal design evaluation for assessing the performance of the built environment. In: Craddock G, Doran C, McNutt L, Rice D (eds) Transforming our world through design, diversity and education: proceedings of universal design and higher education in transformation congress 2018. Studies in Health Technology and Informatics, vol 256, pp 771–779. https://doi.org/10.3233/978-1-61499-923-2-771.
15. Brambilla A, Buffoli M, Capolongo S (2019) Measuring hospital qualities. A preliminary investigation on health impact assessment possibilities for evaluating complex buildings. Acta bio-medica : Atenei Parmensis 90(9S):54–63. https://doi.org/10.23750/abm.v90i9-S.8713
16. Faroldi E, Fabi V, Vettori MP, Gola M, Brambilla A, Capolongo S (2019) Health tourism and thermal heritage. Assessing Italian Spas with innovative multidisciplinary tools. Tour Anal 24(3):405–19. https://doi.org/10.3727/108354219X15511865533121
17. Italian Ministry of Environment, Sea and Land Protection (2013) Green infrastructures and ecosystems services as instruments for environmental policy and green economy: Potentiality, criticality and recommendations. Conference "Nature of Italy" Rome, December 11 and 12, 2013. http://www.minambiente.it/sites/default/files/archivio/allegati/natura_italia/natura_italia_documento_sintesi_finale_eng.pdf
18. European Commission. Ecosystem services and green infrastructure. http://ec.europa.eu/environment/nature/ecosystems/index_en.htm
19. Rojas-Rueda D, Nieuwenhuijsen MJ, Gascon M, Perez-Leon D, Mudu P (2019) Green spaces and mortality: a systematic review and meta-analysis of cohort studies. Lancet Planet Health 3:469–77

Active Cities & Health: A Children Perspective

Antonio Borgogni and Elena Dorato

1 The Active Cities Approach

As a reaction to car-centered urban and mobility planning, and to a related overall decline in environmental sustainability and urban safety, especially during the 1970s several civil movements flourished in central and northern Europe. These were mainly targeting socio-educational, environmental, and town planning issues, shifting the focus from traffic decongestion to urban populations' wellbeing, improving urban renewal operations also through participatory processes. Such experiences focused primarily on traffic calming interventions, with the aim of reducing vehicular traffic flows and speed, while giving the streets and public spaces back to the citizens. Such a dynamic reflects the idea of the so-called "reconquered city" introduced by Danish planner Jan Gehl, referring to these very years as the forefathers of a new trend towards winning back the public spaces [1]. The Dutch *woonerf* (1976) were undoubtedly the first legislative provision going in this direction, standing out for their social and educational implications even before the planning ones [2]. Similar interventions were then realized in Denmark, Germany, Austria, Switzerland, France, and United Kingdom, often known with the name of *Living Streets*, forerunners of today's car-free neighborhoods: spaces where kids could play safely on the streets, and where social relations reappeared as a key-element for urban livability [3]. In Italy, these interventions began to be applied only during the 1990s especially in

---

This paper is the result of a shared work among the authors. In particular A. Borgogni coauthored paragraph 1 and authored paragraph 3; Elena Dorato coauthored paragraph 1 and authored paragraph 2. The authors jointly revised the drafts of the chapter.

A. Borgogni (✉)
Department of Human and Social Sciences, University of Bergamo, Bergamo, Italy
e-mail: antonio.borgogni@unibg.it

E. Dorato
Department of Architecture, University of Ferrara, Ferrara, Italy
e-mail: drtlne@unife.it

the northern regions, and only in recent years studies are focusing on this topic in relation with public health [4, 5]. Starting from the Nineties, participatory actions involving children were carried out for the planning of urban green spaces, safe routes to school, and schoolyards. A prominent even if disregarded role was played by the concept of the body regaining the public space for playing, commuting, walking, cycling within the urban environments [6, 7]. The main aim of these new processes and interventions was to give back to children the space for playing, as well as the freedom of roaming around the neighborhoods.

The challenging times we are currently witnessing because of the Covid-19 pandemic are in a way bringing us back to those reflections and first project implementations, teaching us to share different proximities and to deeply understand—by subtraction—the essentiality of urban public spaces while experiencing, at the same time, the desperate need of our bodies to freely move around the city [8]. Especially in the Mediterranean urban culture, the public dimension of the city has played a fundamental role in defining people's quality of life, often representing actual "extensions" of private homes, with streets being the courtyards of houses, parks and gardens used as playrooms, squares as living-rooms. It is this collective, shared and democratic relevance of the city that—if and when it is lacking—forces us even more to confront the spatial and social inequalities that manifest themselves with increasing intensity, especially among the most vulnerable population.

In this perspective, the emergent *Active Cities* approach has been firstly promoted by the public health sector [9], and investigated through town planning [7], socio-educational, and physical activity perspectives [10]. The key point of the approach is to enhance the opportunities for all citizens to be physically active within their daily routines. The original aim of the World Health Organization was to promote active lifestyles in the urban environment to tackle inactivity-related health issues like non-communicable-diseases, yet understanding that such development should encompass infrastructural, social, educational, and mobility policies and actions. Within this process of growing awareness of the health benefits potentially coming from the city context and its characteristics, the involvement, and the role of physical activity in the urban planning and public health framework also begins to unfold, becoming more and more relevant [7]. It is fundamental to stress that, when addressing physical activity, this work refers to the definition given by the World Health Organization as any bodily movement produced by skeletal muscles that requires energy expenditure; "Physical activity includes recreational or leisure-time physical activity, transportation (e.g. walking or cycling), occupational (i.e. work), household chores, play, games, sports or planned exercise, in the context of daily, family, and community activities" [11, p. 8].

Despite some scattered interventions, in many European cities, frailer groups of citizens from the independent mobility viewpoint, such as children, elderly, and disabled people, are still encountering difficulties in moving autonomously. The lack of autonomy affects people's health, directly reducing their opportunities to learn, socialize and to be physically active in both the formal and informal contexts offered by the urban public realm [12]. Within such framework, the decline of children's autonomy is a recognized concern. The comparative research Children's Independent

Mobility [13], carried out in sixteen countries involving children aged 7–15, ranks Finland first, followed by Germany, Norway, and Sweden, while Italy and Portugal are the two European countries in which children are less autonomous. Overall, Italian children are about three to four years behind the best-ranking countries in their freedom to be independent in several kinds of mobility. More precisely, children's autonomy in walking to school in Italy (7%) is much lower than in England (41%) and Germany (40%) [14]. The Italian situation is, somehow, unique: in fact, due to legal restrictions it is not permitted to a child under the age 14 to roam independently. This normative has been incorporated in the primary school regulations, leading to an overall prohibition to exit school without being picked up by an adult.[1] Playing, walking and cycling independently seem to be unperceived rights; children's rights that adults are not accustomed to respect [15–17].

2 Children in Cities: A Complex Relationship

The fundamental role of children's independent mobility and, more generally, the capability of performing independent activities within the urban context, represents not only a right, but also a great opportunity to develop new, more sustainable, livable and healthy urban environments for all [8]. Recalling the powerful statement of English urbanist-architect Colin Ward: "*I don't want a Childhood City. I want a city where children live in the same world as I do*" [18, p. 204], we could assume that accessibility, safety and comfort are key-characteristics that every urban environment should provide, allowing children to live and experience the city alongside other generations, invoking the urgency of rethinking and updating the city-children dichotomy by placing them at the center of urban policy and design. Even though in a simplified way, these fundamental features of the urban realm somehow represent what Henri Lefebvre defined as the *Right to the City*, including the "*right to freedom, to individualization, to habitat and to inhabit*" [19, p. 173], as well as the right to participation and appropriation of the urban public spaces [20].

As increasingly shown by the international scientific literature, active mobility (i.e. home-to-school journeys on foot, by bicycle, using skates, etc.) and the performance of extra-school time activities (playing or practicing any other kind of PA within the urban environment) represent the main occasions for children to keep physically active within their daily routines, while acquiring new autonomy skills and enhancing social capabilities and networks [16, 21–23]. However, the most recent alarming data describing Italian children's conditions in relation to average levels of practiced

---

[1] According to the Art. 591 of the Penal Code about children neglect; the sentence of Bologna State Legal Advisory Service (2001); the sentence of the Court of Appeal (Corte di Cassazione) n. 21593/2017.

physical activity[2] and autonomous mobility are stressing the need for new inter-disciplinary efforts tackling such inactivity pandemic. In this perspective, today more than ever there is a need for a closer and more efficient collaboration between the fields of urbanism, pedagogy, and physical activity sciences [25].

As discussed by landscape architect Anna Lambertini, over the past two centuries to each social crisis, critical revision of urban transformation, or redefinition of the role of public spaces corresponded a "(…) *Renewed revolt of the homo ludens and the need to reaffirm the essential role of play in the daily lives of adults and children*" [26, p. 11]. Recalling the 1939 well-known essay by Johan Huizinga, arguing that *Homo Ludens* represents an essential function of human existence as much as *Homo Faber*, the action of playing—in its wider meaning—is extremely relevant in forming individuals and building a society. And yet today, for the digital natives the simple act of playing or that of walking to school shifted from a collective to an almost entirely individual activity, from an active to a sedentary one; and from an outdoor to an indoor setting [27].

In his renowned book The Child in the City, Colin Ward [28] already discussed the need of rethinking the child-city dichotomy, placing kids at the center of planning policies and urban design interventions. Children have the right of roaming around the city, and yet urban hostility and inadequacy force them to stay in determined spaces such as the houses, the schoolyards, the sport facilities or other gated gardens and playgrounds. Such changes, besides reflecting a major social crisis and a norma-tive challenge, are also strongly connected to the perception and use of the urban environment not only by children, but also by their parents, whose lifestyles and mobility/activity habits strongly influence and impact those of younger generations.

Thus today more than ever, our interdisciplinary reflections should ground on that same question that, exactly forty years ago, French landscape architects Jean Simon and Marguerite Rouard raised in their famous book Children's Play Spaces: *"The most important question is: are parents, teachers, designers, architects, recreational counsellors, and city planners willing to take the real needs of a children into consid-eration in the cities of the future or in urban renewal projects preserving the cities of the past?"* [29].

As decades went by, unfortunately it seems we are still not able to give an affir-mative answer to such complex and yet elementary question, stressing the need for more effective collaborations among different disciplines, while overcoming the many restrictive regulations and normative issues. However, although still lagging behind other European countries, also in Italy—which had experienced a particularly successful era during the 1990s in relation to children's role within the public realm and to their participation's right—some interesting and potentially relevant projects and researches have recently been developed [17, 25], especially focusing on the role of school courtyards and other school-related outdoor spaces, intended as safe, protected and dedicated areas for kids to play and keep physically and socially active.

---

[2] Recent, still unpublished, researches are supporting that physical inactivity in Italian children is leading to severe motor impediments, and overall changes in body structure, showing worsening data compared to a decade ago [24].

According to the US movement *No Child Left Inside*, especially in the western urban culture, pupils are increasingly growing unaware of environmental issues and culture, developing a nature-deficit-disorder (directly linked to the dependance of our society on digital devices) which affects children's quality of life at many different levels [30], also contributing to the overall sedentary pandemic among the youngsters, while creating psychological and biological pathologies in the development of the child, such as attention deficit, obesity, vision problems, depression [31]. Echoing the aims and critical perspective of the American movement, in the past years also in Italy many non-profit organizations launched a series of awareness raising campaigns focusing on the habits of children during school recess hours,[3] showing some quite alarming data.

Given the seriousness of the matter, there has also been an attempt to intervene from a political point of view,[4] yet failing to be incisive on a national level. A similar trajectory has been undertaken in the United States, with only a few States endorsing the *No Child Left Inside* movement by implementing bills and creating institutional programs in local parks and schools supporting open-air recreational activities. In Italy, only a few cities among which the municipality of Turin have been promoting, in the past years, a virtuous cooperation between educational institutions and the city, considering the schoolyards of the city also as a physical opportunity for children-friendly urban public spaces.[5] In the summer of 2020, especially due to the current Coronavirus pandemic and the great challenges it raises also in terms of rethinking basic public services (and rights) such as the school, the Italian Ministry of Education has drawn up a "Document for the planning of school, educational and training activities", introducing the "Community Educational Pacts": a tool aimed at strengthening territorial alliances to encourage the provision of other structures or spaces, such as parks, theaters, libraries, archives, cinemas, museums, to carry out complementary educational activities, thus physically and metaphorically "expanding" school spaces while supporting the education facilities. Also, the so-called "Simplifications Decree" (Law n.120 September 11, 2020) finally introduced the concept of "school zone", meaning an urban area near school buildings where "special protection of pedestrians and the environment is guaranteed".

Thus, granting children—and consequently a much broader share of the population—the right to safely and autonomously experience the urban city spaces becomes,

---

[3] In 2014, Lipu-BirdLife Italy conducted across the country a research named "Scuole Verdi" addressing several hundreds of teachers and parents, qualitatively investigating children's habits. The emerging data is alarming: only 41.6% of children play outside once a week at most, while 20.5% three to two days a week; 42.2% of interviewed teachers affirmed that children spend an average of only ten hours per month playing in the schoolyard.

[4] In July 2018, Democratic Senator Monica Cirinnà presented a draft law named "Norms for the development of green spaces in school buildings" (Parliament Act n.703), advocating the realization or valorization of green open spaces within the premises of local schools.

[5] Since 2012, often as a result of participatory processes, many of the over 200 schoolyards in Turin have been transformed into more pleasant and activity-conducive environments, open to the city also during extra-school hours. See the Regulation for Managing Activities in Municipal Schoolyards n.359, City of Turin (approved on Nov 26th 2012, implemented from Dec 10th 2012 http://www.comune.torino.it/regolamenti/359/359.htm).

today more than ever, a fundamental goal to be set high on the urban agendas, tackling not only urgent health-related matters, but also cultural and socio-educational issues. As accurately expressed by Giancarlo De Carlo in a 1947 monographic issue of the magazine for architecture DOMUS, in the "stratified city" the "problem of life in the school extends to the problem of life in the city"; and again, "the planning problem of the school has now become the planning problem of the city". Decades before him, eminent educator Maria Montessori had understood the special link between childhood and outdoor activities, grasping the immense educational potential of nature. In *The Montessori Method*, she dedicates an entire chapter to "Nature in Education", considering it one of the most important elements to be integrated within schools; in essence, for Montessori a school without outdoor green space could not exist. Resuming these concepts, Rouard and Simon wrote: "In addition to the common schoolyard, school, of course, should have fields for sports and open-air activities. The banks of a river or stream passing through town may provide useful play space and can be made suitable for that purpose at small cost" [29].

3   The Walk-To-School Research

So far, in Italy, only few researches on Active Cities have been carried out, addressing the previously introduced complex child-city dichotomy. The overall approach on Active Cities has been supported since 2008 by an on-going study based on direct observations (n = 45) of European sites (n = 25) and several interviews (n = 22) [6]. Concerning children, a specific research has been developed in the city of Cassino (Lazio Region, Italy), based on a walk-to-school intervention known as *Pedibus*. The research grounded on the hypothesis that the activation and the implementation of a walk-to-school program, together with training activities for teachers and aware-raising interventions for parents, could positively influence children active mobility and lifestyles [8, 32].

As already mentioned, Italian children are less autonomous than their European mates; in the age 8–11 only 28% (aged 8–11) are active against 40% of French and 52% of Spanish on the route to school. These data are influencing the low rate of moderate and vigorous PA among the Italian 11 years old children and aerobic PA among adolescents [33, 34]. Thus, how could a community-based action promoted by the University and involving children, parents, teachers, local associations, and the municipality influence children's behaviors in a setting that, as showed in previous researches, is strongly oriented towards inactive lifestyles and children's dependency [35, 36]?

The three-year longitudinal research was based on a mixed-method approach [35], involving pupils attending the three public primary schools in Cassino, as well as their parents. A questionnaire on children's autonomy (validated by the Italian National Research Center, CNR-Institute of Cognitive Sciences and Technologies) was used as research tool, submitted before and after the intervention to third, fourth, and fifth graders aged 8–11 (average age 9.2). The students also received a questionnaire addressing their parents: in relation to the three years of the study (2015/2016/2017), a number of 693/741/528 questionnaires were returned by children, and 574/597/422 by parents (only two schools in 2017).

The questionnaire was composed by two main sections: socio-demographic data, and children's autonomy and independent mobility. The questionnaire addressing parents was integrated by sections on sport participation and the use of ICT-devices and internet. Focus groups (n = 09) have also been carried out with teachers (n = 2), parents (n = 2) and pupils (n = 5). Direct observations (n = 8) in the area around the intervention school were carried out during school entrance and exit times. The intervention was firstly centered on a *Pedibus* in two out of three schools (2015, following previous experiences). During the second year of implementation, the action became more frequent towards the end of the school year. In the third year, since spring 2017, the action was carried out once a week in the intervention school for all participating students (average attendance n = 83).

According to the results from the parents' pre-intervention questionnaire, 75.3% of children go to school by car, 7.4% by school-bus, and 17.3% in an active way. Only 3.4% go to school independently. The main motivations preventing families to allow children's independent mobility are: distance (55.3%), traffic dangers (17.8%), and "stranger danger" (15.6%). In extra-school hours, 26.6% use the bicycle near the house, and 12.8% go to friends' homes alone. Focus groups with teachers highlighted the impact of normative restrictions and parents' over-control in the decrease of children's autonomy. Focus groups with parents also stressed the influence of dangers overrepresentation especially triggered by media. School settings and the lack of community-based planning seem, ultimately, to influence in a negative way children independent mobility and their overall autonomy.

Post-intervention results showed a slight increase of active mobility (1%/3% to school/return home) in the intervention school, and a decrease (2%/3% to school/return home) in the control school [32] (Table 1).

The interpretation and discussion of these results opened the way to the development of a conceptual model aiming to embrace the different typologies of physical activity performed by children, considered from the point of view of their independent mobility and autonomy. If looking to the health outcomes, walking to school, playing outside, going autonomously to meet friends or running small errands within the urban public realm represent some routine actions greatly contributing to reach physical activity recommendations. Focusing on education, the same activities are

Table 1 Percentage of children being active: (a) going to school; (b) going back home; in (c) intervention; (d) control schools. Data pre-post walk-to-school intervention; parents' questionnaire [32]

Active mobility	Going to school %		Back home %	
Data	Pre	Post	Pre	Post
Intervention school	17.07	17.89	19.51	22.76
Control school	23.21	21.43	26.79	23.21

crucial to learn competences and the written and unwritten intrinsic rules of the urban environment. Thinking at social aspects, they also allow children to create acquaintanceships and friendships, building relationships without any adult supervision. Considering the *Ecological Model of Health Behaviors* [37], it is evident that organized physical activity or sport are merely included in the "Active recreation and Occupational activities" domains and, within them, few are the settings in which educators or coaches lead or train groups or individuals (physical education classes, walk-to-school programs, sport, fitness courses). Therefore, there seem to be plenty of available time—mostly for commuting, leisure or play—and venues (predominantly urban public spaces) in which to deploy non-organized or non-supervised physical activities. Regarding children, such times and spaces are strictly linked together with their autonomy and, in particular, with their independent mobility and roaming possibilities.

When educating to active lifestyles, the subtraction of the physical playfield entails to play an "impossible game" denying the educational significance of the urban public domain, and that of risk-taking as a pedagogic *dispositif* [38, 39], eventually leading to the exclusion and disappearance of the child from public spaces. Originating from the discussion of the results, and from an interdisciplinary review carried out in the fields of physical activity, pedagogy, public health, town and mobility planning, environmental psychology, and urban sociology, a conceptual model (Fig. 1) was drawn based on the classification of physical activity as independent or non-independent [40]. The first group of physically active behaviors is associated with children's autonomy to roam in the public spaces; the second one assembles activities performed when escorted by adults for mobility or leisure purposes; the third one relates with sport, leisure or educational activities, during and after school time, organized and taught by adults.

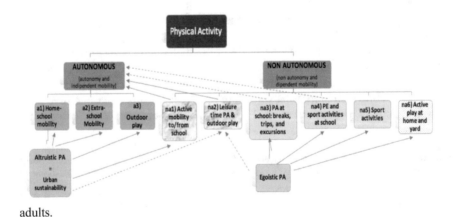

adults.

Fig. 1 Conceptual model of children physical activity [40]

According to the model, a large part of physical activity in children should be considered an epiphenomenon of their independent mobility and mobility opportunities in the urban public spaces. Consequently, to enhance overall children's *activeness* from a decision-making point of view, the focus should be directed towards all the "realms" of experience of the child, also considering the cultural, educational, and legislative determinants of physical activity. Referring to the conceptual model, the first two groups of "autonomous physical activity" address adults' lifestyles and the urban planning and mobility culture, while the others represent a matter of organization—including spatial issues—in which children are confined into specific spaces, more or less attractive and designed for them. Within such framework, two main questions arise: do we really think that the majority of children could be more active in the settings of the model's third group? And also, is it plausible, from an organizational (e.g. family, school, clubs) and economical point of view that a preponderance of children participates in sport activities to achieve the 60 min of moderate or vigorous physical activity, as recommended by the WHO as minimum target to stay healthy?

Certainly, it is possible and desirable to participate more in sport, and to increase physical education in schools, together with more active recesses and excursions. Nevertheless, it is out of the urban confined spaces and times where we can play a "possible game", or rather in the vast realm of the children conceivable experiences. Therefore, to enhance children physical activity within the urban environment a shift in the prevailing adult's mentality is needed. Such change involves many interrelated aspects, but the infrastructural one is the prerequisite having the major influence on behaviors [41].

In this perspective, also a radical change in town and mobility planning is required, at the different architectural scales: from healthy and sustainable building design; to the conception of active and easily accessible public spaces, playgrounds and school-yards; to the realization of more widespread and safe active mobility infrastructures, efficiently connecting the public realm and allowing all citizens—children first—to autonomously move around the city on an "altruistic" way (referring to the model) substituting active mobility to the motorized one instead of only practicing physical exercise or sport ("egoistic" way). Rephrasing Ward, while looking at enhancing health through everyday physical activity: we do not need cities, districts, spaces designed for determined categories of people, nor for the practice of specific sport or activities preventing the coexistence of many other uses and users. Today more than ever, we do need cities in which everyone's body practices are facilitated by a safe, attractive, connected, and vibrant *milieu*. Unquestionably, in such an environment, children are those who might get the more important benefits in term of establishing enduring active lifestyles.

References

1. Gehl J, Gemzoe L (2006) New city spaces. The Danish Architectural Press, Copenhagen

2. Borgogni A, Dorato E (2020) Ripensare l'urbanità dell'urbano. Dalla strada alle strade. In: Ceruti M, Mannese E (eds) Racconti dallo Spazio. Per una pedagogia dei luoghi. Pensa MultiMedia, Lecce, pp 41–75
3. Dorato E (2015) La Città Attiva. Nuovi approcci al progetto degli spazi pubblici urbani/The Active City. New approaches to the design of urban public spaces. Paesaggio Urbano 1:52–57
4. WHO Europe (2017) Towards more physical activity in cities. WHO Europe, Copenhagen
5. Capolongo S et al (2018) Healthy design and urban planning strategies, actions, and policy to achieve salutogenic cities. Int J Environ Res Public Health 15(12):2698. https://doi.org/10.3390/ijerph15122698
6. Borgogni A (2012) Body, town planning, and participation. the roles of young people and sport. Jyväskylä University Printing House, Jyväskylä
7. Dorato E (2020) Preventive urbanism. The role of health in designing active cities. Quodlibet, Macerata
8. Dorato E, Borgogni A (2020) Active cities/active children: a planning and pedagogical perspective. Convergências - Revista de Investigação e Ensino das Artes XIII(26)
9. Edwards P, Tsouros AD – WHO (2008) A healthy city is an active city, a physical activity planning guide. WHO Europe, Copenhagen
10. Borgogni A, Farinella R (2017) Le città attive. Percorsi pubblici nel corpo urbano. Franco Angeli, Milan
11. WHO (2010) Global recommendations on physical activity for health. WHO Europe, Geneva
12. Borgogni A, Dorato E, Arduini M, Ciccarelli M, Digennaro S, Farinella R (2017) Active and sociable cities: the frailer elderly's perspective. In: Caudo G, Hetman J, Metta A (eds) Compresenze. Corpi, azioni e spazi ibridi nella città contemporanea. RomaTrePress, Rome, pp 45–47
13. Shaw B, Bicket M, Elliott B, Fagan-Watson B, Mocca E, Hillman M (2015) Children's independent mobility: an international comparison and recommendations for action. Policy Studies Institute, London. http://westminsterresearch.wmin.ac.uk/15650/1/PSI_Finalreport_2015.pdf. Last Accessed 31 Jan 2018
14. Renzi D, Prisco A, Tonucci F (2014) L'autonomia di movimento dei bambini: una necessità per loro, una risorsa per la scuola e per la città. Studium Educationis 15(3):105–119
15. Gray P (2011) The decline of play and the rise of psychopathology in children and adolescents. Am J Play 3(4):443–463
16. Mackett RL (2013) Children's travel behaviour and its health implications. Transp Policy 26:66–72
17. Borgogni A, Arduini M (2017) Le città sostenibili dei bambini: la progettazione partecipata degli spazi urbani. In: Birbes C (ed) Trame di sostenibilità. Pedagogia dell'ambiente, sviluppo umano, responsabilità sociale. Pensa MultiMedia, Lecce, pp 183–198
18. Ward C (1978) The child in the city. Pantheon Books, New York City
19. Lefebvre H (1996) Writings on cities (trans and eds: Kofman E, Lebas E). Blackwell, Oxford
20. Franck KA, Stevens Q (2007) Loose space. possibility and diversity in urban life. Routledge, Abingdon
21. Grize L, Bringolf-Isler B, Martin E, Braun-Fahrländer C (2010) Research Trend in active transportation to school among Swiss school children and its associated factors: three cross-sectional surveys 1994, 2000 and 2005. Int J Behav Nutr Phys Act 28(7):1–8
22. Mitra R (2013) Independent mobility and mode choice for school transportation: a review and framework for future research. Transp Rev 33(1):21–43
23. Ikeda E, Hinckson E, Witten K, Smith M (2019) Assessment of direct and indirect associations between children active school travel and environmental, household and child factors using structural equation modelling. Int J Behav Nutr Phys Act 16:32
24. Filippone B, Vantini C, Bellucci M, Faigenbaum AD, Casella R, Pesce C (2007) Trend secolari di involuzione delle capacità motorie in età scolare. Studio longitudinale su un campione regionale italiano. SdS Rivista di Cultura Sportiva 72:31–41
25. Borgogni A, Arduini M, Dorato E (2018) Le città sostenibili dei bambini: sfide e opportunità per un'urbanistica democratica. In: Proceedings of the XX SIU national conference "Urbanistica

è/e azione pubblica. La responsabilità della proposta". Planum Publisher, Rome-Milan, pp 1604–1608

26. Lambertini A (2017) Editorial: let's play. Architettura del Paesaggio – Playtimes 35(2):10–12
27. Covolo L, Ceretti E, Moneda M, Castaldi S, Gelatti U (2017) Does evidence support the use of mobile phone apps as a driver for promoting healthy lifestyles from a public health perspective? A systematic review of randomized control trials. Patient Educ Couns 100(12):2231–2243
28. Ward C (1977) The child in the city. The Architectural Press Ltd., London
29. Rouard M, Simon J (1977) Children's play spaces. From sandbox to adventure playground. The Overlook Press, Woodstock, NY
30. Louv R (2005) Last child in the woods: saving our children from nature-deficit-disorder. Algonquin Books of Chapel Hill, Chapel Hill NC
31. Quesada D, Ahmed NU, Fennie KP, Gollub EL, Ibrahimou B (2018) A review: associations between attention-deficit/hyperactivity disorder, physical activity, medication use, eating behaviors and obesity in children and adolescents. Arch Psychiatr Nurs 32(3):495–504
32. Arduini M (2018) Mobilità scolastica attiva e autonomia nella scuola primaria. Il ruolo del Pedibus e dei dispositivi portatili nella promozione di stili di vita attivi, Doctoral Dissertation, University of Cassino and Southern Lazio
33. WHO (2016) Growing up unequal: Gender and socioeconomic differences in young people's health and well-being. HBSC study: International Report from the 2013–14 survey. WHO Europe, Copenhagen
34. Eurostat (2017) Time spent on health-enhancing (non-work-related) aerobic physical activity by sex, age and income quintile. Eurostat dataset. http://ec.europa.eu/eurostat/web/products-datasets/-/hlth_ehis_pe2i. Last Accessed 20 Sept 2020
35. Pompili L. Borgogni A (2013) The influence of the playgrounds georeferencing, layout, and equipment in the development of active life styles in children. Sport Sci Health 9(1):p. addenda
36. Arduini M, Borgogni A, Capelli G (2016) The walk to school actions and portable devices as means to promote children's active lifestyle. In: Novak D, Antala B, Knjaz D (eds) Physical education and new technologies. Croatian Kinesiology Association, Zagreb, pp 19–25
37. Sallis JF, Cervero RB, Ascher W, Henderson KA, Kraft MK, Kerr J (2006) An ecological approach to creating more physically active communities. Ann Rev Public Health 27:297–322
38. Massa R (1989) Linee di fuga. L'avventura nella formazione umana. La Nuova Italia, Florence
39. Farné R, Agostini F (2014) Outdoor education. L'educazione si-cura all'aperto. Edizioni Junior, Parma
40. Borgogni A, Arduini M, Digennaro S (2018) Mobilità attiva, autonomia e processi educativi nell'infanzia e nell'adolescenza, in Metis (Special Issue December 2018):33–45
41. Sallis JF, Cerin E, Conway TL, Adams MA, Frank LD, Pratt M, … Davey R (2016) Physical activity in relation to urban environments in 14 cities worldwide: a cross-sectional study. The Lancet 387(10034):2207–2217

Design for All: Strategy to Achieve Inclusive and Healthier Environments

Erica Isa Mosca

1 Introduction

Health is a value currently associated to a social model, that results from social, cultural and economic factors, as well as environmental [1]. The International Classification of Functioning, Disability and Health (ICF) [2] indicates health and disability as conditions reliant by a plurality of factors, including Environmental and Personal factors. Indeed, different scientific studies demonstrate that design features can influence people behavior and well-being [3–5]. In particular Evidence-Based Design (EBD) methodologies prove that the physical environment can have impacts on users' health related outcomes and performance (e.g. reducing staff stress and fatigue, improving patient safety, reducing family stress, improving communication between staff and patient and overall satisfaction) [6–8]. This is especially true in in healthcare facilities as a place where physical, perceptive, cognitive and social aspects should be considered in the design as main priorities, since people and environment influence each other [9, 10].

Although different factors can influence users' well-being, health and safety, architectural environments often have not been designed considering the people's perception, usability and experience within the space [11]. This can compromise the performance of the entire service (e.g. disorientating layout, unwelcoming environments, unclear information, unpractical healthcare settings, etc.) [12]. Indeed, when social aspects are left at the end of the design process, there are no tangible results on individuals well-being and it can even generates extra cost with time waste and disabling situations [11]. Hence, a more holistic approach is needed to integrate the human factors to the design process providing a real impact on the healing performance of the buildings, linking different skills and needs [13].

E. I. Mosca (✉)

Department of Architecture, Built environment and Construction engineering (ABC), Politecnico Di Milano, Milan, Italy

e-mail: ericaisa.mosca@polimi.it

S. Capolongo et al. (eds.), *Therapeutic Landscape Design*,
PoliMI SpringerBriefs, https://doi.org/10.1007/978-3-031-09439-2_11

In this regard, Design for All (DfA) strategy, was defined to satisfy the unexpressed needs of the greatest number of people, derived by the society transformation over the last decade (e.g. growing of aging level and people with impairments) [14, 15]. The goal of this article is to explore how DfA can be applied into the design process, in relation to goals aimed at satisfying different users' needs. Attention will be placed on healthcare facilities where a plurality of users with different needs interact in these spaces (patients, clinicians, technicians, visitors, etc.), characterized by a number of functions being carried out in the same location [11, 12]. Two different case studies of healthcare facilities will be analyzed to highlight both the *design process* and *design solutions* in relation to DfA. The article expresses how this strategy can be used to enhance inclusive design of buildings, providing positive outcomes on users such as usability, well-being and social inclusion.

2 Design for All Strategy

DfA is a design strategy, which aims to social inclusion and uses' well-being. This strategy presents a more holistic approach thinking about individuals, which are considered in the design project for their needs and wishes, instead for their abilities or disabilities. The official definition of DfA was provided in the Stockholm Declaration as the 'design for human diversity, social inclusion and equality' [14] by the EIDD—Design for All Europe, a European network, founded in Dublin by the designer Paul Hogan in 1993 with the name of European Institute for Design and Disability.

The evolution of disability concept contributes to define DfA strategy. Indeed, disability is currently defined by the ICF [2] as the interaction between different factors: body (function and structure), activity, participation and context (environmental and personal factors). This means that it refers to a universal human experience since all individuals can be permanent impaired (e.g. people born blind), temporary impaired (e.g. age, pregnancy, broken limb) or situational impaired, considering the relation among health condition, built environment and social factors (e.g. negative attitudes, inaccessible transportation and public buildings, and limited social supports) [16]. Starting from this assumption, DfA strategy shifts the concept of 'disability' from a *medical model*, which focuses on 'special needs' of categories of people with disabilities or impairment, to a *cultural and social model,* in which the attention is focused on all people needs [17] and people with impairments are considered as experts, with knowledge about disabling and enabling environments [18]. In this regard, the experience of users is meant as not just related to physical or perceptive conditions (accessibility), but also cognitive, sensory, and social ones for ensuring their well-being [19].

3 Achieving Inclusive and Healthier Environment Through Design for All Strategy

The following paragraphs describe how to integrate DfA within the design process and solutions for achieving inclusive environments.

3.1 Design Process: Users' Involvement

The Stockholm Declaration specifies that DfA "requires the involvement of end users at every stages of the design process" [14]. The active involvement of final users is one of the key aspects applying DfA strategy. Users' needs and wishes should be considered in designing and evaluating buildings from the beginning of the design process, in order to prevent further changes [19, 20]. Indeed, when social aspects and usability problems are addressed after construction, there are no tangible results on individuals' well-being, furthermore it is time consuming and it can even generate disabling situations and extra-costs [21]. Following this perspective, DfA suggests that the design project development should be participatory, through a dialogue among actors, from the decision-makers to the final users [22]. All the experts in each discipline should be involved at each phase of the process (e.g. understanding of needs, design, evaluation, business-manufacturing). For instance, in healthcare environments clinicians might lead on the initial clinical research, but all other members of the team (design, business, administrative ones, etc.) take part in the process [23]. At the same time, the final users are considered as experts, because their experience is crucial to know their needs and wishes, to be transformed into actual design solutions (i.e. perception of the space by blind people) [18]. Therefore, designers are invited to consider multisensory experience of different users to improve the quality of the space.

3.2 Design Process: Users' Activities and Context

Both ICF model [2] and Ergonomics [14, 17] consider that the psychophysical characteristics of users are strictly related to the environment. Starting from this assumption, DfA adopts the same concept in a larger scale of the human diversity. Indeed different studies on DfA take into account the relation between human *activities* (e.g. overcome distance, orientate, concentrate, relaxing, etc.) and potential *users' characteristics* (e.g. age, gender, culture, abilities, disabilities, etc.) in different circumstances of the *built environment* (e.g. outdoor spaces, entrance, horizontal circulation, rooms, toilets etc.) [21, 22, 24–26]. In this regard, actions of people and their needs are constantly placed in relation to the context, representing the socio-spatial backdrop in which an activity may occur. They are also called 'activity settings' [27] or 'stages of a travel chain' [24], which usually are identified as: arriving at or approaching a building; entering a building; circulating through the building; interacting with the main building facilities. In line with this, Luigi Bandini Buti has described how DfA was introduces for designing a refreshment area, considering different circumstances of the space and related users' activities: approaching (choosing the facility, parking, reaching the entrance); entrance/welcoming (using doors, wishes of the users, how to choose in food facilities); using restaurants and bars (eating and drinking, using seats and tables, garbage collection); etc. [26].

In the context of healthcare facilities, where tasks considerably vary in relation to different spaces [12], DfA strategy can be integrated to improve human experience of users. For instance, in waiting rooms it is fundamental to consider different aspects of the built environment related to the action performed in that space: how the patients will be greeted; which will be their first view; how patients and visitors will spend time during waiting; how will be the environmental conditions perceived by people waiting (e.g. temperature, acoustic, light, etc.); how will be the admission and administration procedures; which kind of seating can be more comfortable during waiting to fit different users' needs [28]. In this regard, the image shows the variables to consider during the design process following DfA strategy, structured on healthcare facilities design, such as hospitals (Fig. 1). In the case studies' descriptions, environments and users' activities will be constantly related to their needs.

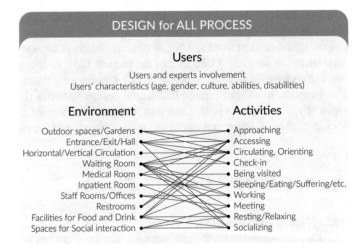

Fig. 1 Variables to consider in design process according to Design for All strategy: users' involvement and characteristics; environmental elements; human activities. *Source* authors. Study part of the doctoral research work in progress by Erica Isa Mosca at Politecnico di Milano entitled "Design for All AUDIT (Assessment Usability & Inclusion Tool). A performance-based tool to evaluate quality of healthcare environments for different users"

3.3 Design Solutions: Principles and Goals of Universal Design

While DfA focuses more on the design process, theoretical basis for the application of the design strategy are given by Universal Design (UD) [29, 30], which is a discipline developed in USA, that shares with DfA the same objectives of considering design for and with 'diversity' of users [15]. UD defined *7 Design Principles* to support the practical application of the strategy: Equitable use, Flexibility in use, Simple and Intuitive use, Perceptible Information, Tolerance for Error, Low Physical Effort and Size, and Space for Approach and Use [31]. In this sense, UD provides a more pragmatic and operative support to designer; however the risk is to apply its Principles as guidelines through a prescriptive approach, instead of proposing design solutions based on real people's needs [22]. Therefore, UD Design Principles should be used only together an in-depth analysis of the users' needs, in order to understand in which circumstance they can support the designer.

Steinfeld and Maisel [32] highlighted that UD should take into account also psychological and social elements, over human performance and sensorial factors already included by the UD Principles. In this way, they can cover all the outcomes of the environment on users. For this reason, the Center for Inclusive Design and Environmental Access (IDeA Center) proposed the *8 Universal Design Goals* [32, 33]:

1. Body fit—accommodating a wide range of body sizes and abilities.
2. Comfort—keeping demands within desirable limits of body function and perception.
3. Awareness—ensuring that critical information for use is easily perceived.
4. Understanding—making methods of operation and use intuitive, clear, and unambiguous.
5. Wellness—contributing to health promotion, avoidance of disease, and prevention of injury.
6. Social integration—treating all groups with dignity and respect.
7. Personalization—incorporating opportunities for choice and the expression of individual preferences.
8. Cultural Appropriateness—respecting and reinforcing cultural values and the social and environmental context of any design project.

The UD Goals actively sought to address the intersection of human performance (Goals 1–4) and social participation (Goals 5–8), where wellness (health) represents a bridge goal that address both themes [25].

4   Case Studies: Design for All in Healthcare Design

Two case studies of healthcare facilities, a hospital and a hospital's healing garden
are analyzed to show the application of DfA principles in practice concerning both
the indoor and outdoor environments. In the first one, DfA strategy is applied during
the design of a new building, while in the latter DfA is used to renovate part of the
architectural environment.

The current analysis describes for both projects the way DfA was introduced
from the beginning of the *design process*. In addition, for each case study, DfA
design solutions are described highlighting the relation between goals and healing
outcomes on users (Tables 1 and 2) ensued by the application of the strategy in the
design of the space.

4.1   St. Olavs Hospital

St. Olavs Hospital in Trondheim, Norway, was the winner of both the Innovation
Award for Universal Design in 2014 and of the category for Architecture and for
Landscape. The hospital is characterized by an inclusive architecture that empha-
sizes the human scale and experience from different perspectives, rather than just on

Table 1 Comparison between goals, design solution adopted in the case study and healing
outcomes. St. Olavs hospital

Goals	Design solutions	Healing Outcomes
Cultural appropriateness	In depth analysis of users' needs End-Users survey Experts collaboration	Understanding users' needs in different circumstances
Social integration	Hospital square	Improving sense of community
Body fit	Entrance with no chance in level Entrance integration of ramps and stairs	Decreasing fatigue, Improving usability, functionality and equal use
Comfort Wellness Body fit	Healing garden Training path for wheelchair	Improving perceived quality Improving overall wellbeing
Understanding Awareness Wellness	Corridor's windows	Improving orientation, Improving wellbeing Decreasing stress
	Ward layout	Improving work performance Communication between staff and patients Perceived safety-security, privacy

Table 2 Comparison between goals, design solution adopted in the case study and healing outcomes. Grenville ward garden

Goals	Design solutions	Healing outcomes
Cultural appropriateness	Users involvement in garden decision making process, Experts collaborations	Understanding users' needs in different circumstances
Body fit Wellness	Specialized handrails Zero step entry Flush paved surface Raised planters	Improving usability and equal use Decreasing fatigue Improving performance in gardening activities
Personalisation Social Integration	Flexible seating	Socialization and privacy Improving overall wellbeing
Comfort Wellness	Water wall	Improving overall wellbeing Decreasing stress

functional aspects [34, 35]. St. Olavs Hospital represents an example of the ambitious agreement of the Norwegian Government, that in 2008 sets the goal that Norway should be characterized by universal and accessible design throughout by 2025 [36].

4.1.1 Design Process and Approach

The design's purpose was making the hospital as a social district accessible and open to the city, as a pleasant and welcoming place for patients, relatives, employees, students, scientists and, in general, for the community.

DfA was implemented at St. Olavs Hospital from the beginning of the design process, with guidance and thematic plans for every aspect of the design [34, 35]. In line with the DfA strategy, users' experience was considered fundamental in the decision-making processes from the first design phases. Indeed, designers assumed patients' point of view to understand how the physical surroundings can be designed to enable treatment. Furthermore, both patients and employees were involved through interviews to understand their different needs and perspectives: while patients explained their emotional needs, staff highlighted the functionality demands of their daily activities. In particular, user surveys of patients revealed three main wishes: privacy, visibility and availability of personnel, and accessibility [34, 35, 37].

One of the most important factors for the success of this project was the collaboration and teamwork of different experts across various disciplines, such as designers, engineering, builders and groups of stakeholders. Following this approach, the landscape architects selected the plans in consultation with the Norwegian Asthma and Allergy Association [34].

4.1.2 Design Solutions

The hospital is an attractive gathering place for the citizens and it represents a medical district opens to the neighborhood. At the entrance of the two main buildings a square with benches and green areas (Fig. 2) is used by both the hospital's users and by the citizens, as a real public space of the city. The square supports pedestrian accessibility by slowing down the traffic in front of the hospital, improving walkability and social interaction.

The main entrance is completely accessible with no change in level. In addition, the staircase of the hospital used to access to an historical building, represents DfA application in a creative way, solving a challenging difference in height with a combination of a stair and a ramp that integrates form and function (Fig. 3).

Inside the hospital, the circulation is supported by big windows that provide natural light and open views, guaranteeing both a direct contact with outdoor space and supporting the orientation in corridors (Fig. 3). Furthermore, the wayfinding

Fig. 2 Square of the main entrance of St. Olavs Hospital, Trondheim, Norway. Credits: Erik Børseth, Ingvild Aarseth, St. Olav Hospital, Trond Heggem. Images retrieved from: DOGA website www.inclusivedesign.no

Fig. 3 Entrance of historical building and hall of St. Olavs Hospital, Trondheim, Norway. Credits: Erik Børseth, Ingvild Aarseth, St. Olav Hospital, Trond Heggem. Images retrieved from: DOGA website www.inclusivedesign.no

system uses a color palette associated to the furniture, in order to recognize different spaces (Fig. 3).

Regarding the space for care, during the design process patients highlighted the need of both privacy and available personnel. For this reason, each ward has been given a center with eight single rooms located off this. In this way, the patients sleep peacefully and employees have a better overview, improving the overall security [34].

The hospital includes different healing gardens stimulating the body and the senses. In particular, a training path for wheelchair users was designed for rehabilitation and it gives the opportunity to practice in private and safe surroundings [35]. Healing gardens are also used by patients to spend time with relatives and by the staff for having break, improving the overall satisfaction and wellbeing of the hospital users.

Table 1 highlights the relation among project goals (UD goals), the described DfA design solutions and healing outcomes on users ensued by the application of the strategy.

4.2 Grenville Ward Garden—Royal Cornwall Hospital

Grenville Ward Garden at the Royal Cornwall Hospital (Fig. 4), in Truro, Cornwall, United Kingdom, represents an intervention of renovation for the entire hospital that meets the needs of elderly post-operative patients, staff and visitors [38].

Before the new design proposal, there was no dedicated users and workers rest area available. Thus, the purpose was to create a series of therapeutic and inclusive environments for users, therapists, staff, and visitors (Fig. 4).

 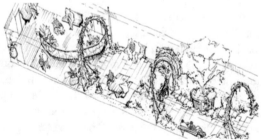

Fig. 4 Grenville Ward Garden Hall and design sketch. Truro, Cornwall, United Kingdom. Credits: Michael Westley, Images retrived from: Westley Design ltd website www.westleydesign.co.uk and https://universaldesigncasestudies.org/

4.2.1   Design Process and Approach

The design aimed to produce a suite of outdoor 'rooms' with varying multi-sensory opportunities for patient interaction, rehabilitation, and recreation appropriate to the elderly users and visitors [38, 39]. The hospital staff also required a rest area with a designated degree of separation from the patient area; hence, the 'room concept' was applied to the overall design [40]. The result is an inclusive and therapeutic garden supporting the rehabilitation purposes of the Grenville Ward Hospital clinical program (for post-operative elderly care), as well as welcoming the wider hospital community.

Working closely with staff rehabilitation specialists, the garden was designed by involving site users, clinicians, and care professionals. Furthermore, the collaboration between the hospital staff and hospital management client team facilitated an inclusive design consultation process [39, 40]. The staff and patients were involved in the garden-making process, they participated in planting workshops and watched the artist produce sculptural elements for the garden on–site, creating a sense of ownership.

The garden was designed to be an inclusive place, taking into account different users' needs. Patients can take part of the welcoming outdoor environment either as an active treatment venue or as a place of respite and general relaxation. They can both enjoy the view from inside or go outside through beds, chairs or using waling aids [39].

4.2.2   Design Solutions

Different experts shared their interdisciplinary knowledge for the design of a range of furnishings to respond to the specific needs of the Grenville Ward. In particular, Westley Design architecture office collaborated with furniture designers, metal artist, hospital arts' coordinator, staff physiotherapists and occupational therapists [40]. Some examples of inclusive design solutions developed to improve the usability are specialized handrails, installed in the garden for use in mobility therapy programs, articulating the subspaces along the length of an otherwise rectilinear courtyard; the design elements are also useful at all times for multiple types of users. In order to promote easy and equal access to the space the pavement has no chance in level and flush paved walking surfaces are used. Raised planters and raised beds are installed throughout the garden for guaranteeing to people with reduced mobility or using wheelchair to interact with plants [38]. Flexible seating options to encouraging socialization or allowing privacy are used and these areas are sheltered with pergolas. Regarding perception's aspects, plantings were chosen to appeal to multiple senses and a water wall, used to separate the garden into staff and patient spaces provide privacy and it can be viewed and listened to by all users ensuring therapeutic effects.

As a result, after completion the use of the garden increased of 150% [40], transforming the paved courtyard, used as a smoking area by staff, to an high quality therapeutic courtyard garden for staff, patients and visitors.

5   Conclusions

The current study described the characteristics of DfA strategy, which can represent a valid proposal to stress the impact on both social aspects and healing performance of the service. This inclusive approach places the experience and perception of the person at the center of the process and it studies their needs and wishes closely. Furthermore, it promotes holistically the participation of different actors (e.g. designers, clinicians, client, workers, final users, etc.) with various knowledge and expertise in the design project.

The study provides descriptive information to integrate DfA considering the management of the design process and practical solutions adopted in relation to goals for satisfying users' needs. Two case studies of healthcare facilities have been used as examples to highlight how Design for All can raise positive outcomes on users through a more inclusive and healthier environment. Further research can follow applying the same method of analysis, in order to collect more evidence about the impact on users' well-being by the application of this strategy.

Thus, as described in the case studies, when DfA is integrated from the first stages of the process, as a care-oriented approach, it can influence on the physical, sensory, cognitive and social well-being for the greatest number of people.

References

1. World Health Organization (WHO) (1946) Preamble to the constitution of the world health organization. Official Records of the World Health Organization. No 2, p 100
2. World Health Organization (2001) International classification of functioning, disability and health: ICF. World Health Organization, Geneva
3. Bitner MJ (1992) Servicescapes: the impact of physical surroundings on customers and employees. J Market 57–71
4. Malkin J (2003) The business case for creating a healing environment. Center for Health Design Business Briefing: Hospital Engineering & Facilities Management, 1
5. Buffoli M, Rebecchi A, Gola M, Favotto A, Procopio GP, Capolongo S (2018) Green soap. A calculation model for improving outdoor air quality in urban contexts and evaluating the benefits to the population's health status. In: Mondini G, Fattinnanzi E, Oppio A, Bottero M, Stanghellini S (eds) Integrated evaluation for the management of contemporary cities. Springer, Green Energy and Technology, pp 453–467. https://doi.org/10.1007/978-3-319-78271-3_36
6. Ulrich RS, Berry LL, Quan X, Parish JT (2010) A conceptual framework for the domain of evidence-based design. HERD: Health Environ Res Design J 4(1):95–114
7. Ulrich RS, Zimring C, Zhu X, DuBose J, Seo H-B, Choi Y-S, Joseph A (2008) A review of the research literature on evidence-based healthcare design. HERD: Health Environ Res Design J 1(3):61–125
8. Brambilla A, Buffoli M, Capolongo S (2019) Measuring hospital qualities. A preliminary investigation on health impact assessment possibilities for evaluating complex buildings. Acta bio-medica: Atenei Parmensis 90(9S):54–63. https://doi.org/10.23750/abm.v90i9-S.8713
9. Capolongo S, Gola M, Brambilla A, Morganti A, Mosca EI, Barach P (2020) COVID-19 and healthcare facilities: a decalogue of design strategies for resilient hospitals. Acta Bio Med. https://doi.org/10.23750/abm.v91i9-S.10117

10. Brambilla A, Capolongo S (2019) Healthy and sustainable hospital evaluation-A review of POE tools for hospital assessment in an evidence-based design framework. Buildings 9(4):76. https://doi.org/10.3390/buildings9040076
11. Buffoli M, Bellini E, Bellagarda A, di Noia M, Nickolova M, Capolongo S (2014) (2014) Listening to people to cure people: the LpCp—tool, an instrument to evaluate hospital humanization. Ann Ig 26(5):447–455. https://doi.org/10.7416/ai.2014
12. Capolongo S, Buffoli M, di Noia M, Gola M, Rostagno M (2015) Current scenario analysis. In: Capolongo S, Bottero MC, Buffoli M, Lettieri E (eds) Improving sustainability during hospital design and operation: a multidisciplinary evaluation tool. Springer, Cham, pp 11–22. https://doi.org/10.1007/978-3-319-14036-0_2
13. Capolongo S (2016) Social aspects and well-being for improving healing processes' effectiveness. Ann Ist Super Sanita 52(1):11–14. https://doi.org/10.4415/ANN_16_01_05
14. EIDD (2004) The EIDD stockholm declaration. http://dfaeurope.eu/wp-content/uploads/2014/05/stockholm-declaration_english.pdf
15. Persson H, Åhman H, Yngling AA, Gulliksen J (2015) Universal design, inclusive design, accessible design, design for all: different concepts—one goal? On the concept of accessibility—historical, methodological and philosophical aspects. Univ Access Inf Soc 505–526 14:505. https://doi.org/10.1007/s10209-014-0358-z
16. Clarkson PJ, Coleman R (2015) History of inclusive design in the UK. Appl Ergon 46 (Part B):235–247
17. Steffan IT, Tosi F (2012) Ergonomics and design for all. Work 41:1374–1380. https://doi.org/10.3233/WOR-2012-0327-1374
18. Herssens J (2013) Design(ing) for more. Towards a global design approach and local methods. Incluside, Asia
19. Mosca EI, Capolongo S (2020) Universal design-based framework to assess usability and inclusion of buildings. In: Gervasi O et al (eds) Computational science and its applications—ICCSA 2020. ICCSA 2020. Lecture Notes in Computer Science, vol 12253. Springer, Cham. https://doi.org/10.1007/978-3-030-58814-4_22
20. Singh R, Tandon P (2018) Framework for improving universal design practice. Product Dev 22(5):377–407
21. Afacan Y, Erbug C (2009) An interdisciplinary heuristic evaluation method for universal building design. Appl Ergon 40:731–744
22. Mosca EI, Herssens J, Rebecchi A, Strickfaden M, Capolongo S (2019) Evaluating a proposed design for all (DfA) manual for architecture. Adv Intell Syst Comput 776:54–64
23. Myerson J, West J (2015) Make it better: how universal design principles can have an impact on healthcare services to improve the patient experience. Universal Design in Education Dublin, Ireland, 12–13 November 2015
24. Froyen H, Verdonck E, De Meester D, Heylighen A (2009) Mapping and documenting conflicts between users and built environments. In: Proceedings of include include 2009. Helen Hamlyn Centre. London
25. Sanford JA (2012) Universal design as a rehabilitation strategy: design for the age. Springer Publishing Company
26. Bandini Buti L (2013) Design for all. Aree di ristoro. Il caso autogrill. Maggioli Editore
27. O Shea EC (2014) Evaluating universal design: exploring methodologies for rating universal design qualities in buildings, [thesis], Trinity College, Dublin, Ireland. Department of Civil, Structural and Environmental Engineering. (UDBRI Tool). http://www.tara.tcd.ie/handle/2262/80435
28. Arneill AB, Devlin AS (2002) Perceived quality of care: The influence of the waiting room environment. J Environ Psychol 22(4):345–360
29. Mace R (1985) Universal design, barrier free environments for everyone. Designers West, Los Angeles
30. Preiser W. F. E, Ostroff E. (2001) Universal Design Handbook. McGraw Hill Professional
31. Connell BR, Jones M, Mace R, Mueller J, Mullick A, Ostroff E, Sanford J et al (1997) The principles of universal design. NC State University, CUD, Raleigh

32. Steinfeld E, Maisel JL (2012) Universal design: creating inclusive environments. John Wiley & Sons, Canada
33. Maisel JL, Ranahan M (2017) Beyond accessibility to universal design. https://www.wbdg.org/design-objectives/accessible/beyond-accessibility-universal-design
34. Design and architecture Norway (2010) Inclusive design—a people centered strategy for innovation. St. Olavs Hospital. Web access March 2019. http://inclusivedesign.no/landscape-architecture/st-olavs-hospital-article175-260.html
35. Eikhaug O, Gheerawo R, Støren Berg M, Plumbe C, Kunur M, Høisæther V (2019) Innovating with people—Inclusive design and architecture
36. Bendixen K, Benktzon M (2013) Design for all in Scandinavia e A strong concept. Appl Ergon 46:248–257. https://doi.org/10.1016/j.apergo.2013.03.004
37. Doezema M (2018) With a deadline in place, Norway warms up to universal design. Web access February 2019 CityLab Website bloomberg.com/news/articles/2018-11-27/inclusive-design-in-norway-s-st-olav-s-hospital
38. Westley Design. Inclusive Design Architecture Website. Web access February 2019. http://www.westleydesign.co.uk/clients-projects/grenville-ward.html
39. Westley Design ltd. (2004) Healing & Care. https://westleydesign.co.uk/healing-care/
40. Institute of Human Centered Design (IHCD).Universal Design Case Studies. Web Access February 2019. https://www.universaldesigncasestudies.org/outdoor-places/parks-gardens/grenville-ward-garden

Printed in the United States
by Baker & Taylor Publisher Services